P9-AZX-254

Moonshot

Moonshot

Inside Pfizer's Nine-Month Race to Make the Impossible Possible

DR. ALBERT BOURLA

HARPER
BUSINESS
An Imprint of HarperCollins*Publishers*

 For information, address HarperCollins Publishers, 195 Broadway, New York, NY 10007.

HarperCollins books may be purchased for educational, business, or sales promotional use. For information, please email the Special Markets Department at SPsales@harpercollins.com.

FIRST EDITION

Designed by Bonni Leon-Berman

Library of Congress Cataloging-in-Publication Data has been applied for.

ISBN 978-0-06-321079-0

22 23 24 25 26 LSC 10 9 8 7 6 5 4 3 2 1

To the more than 5 million individuals
who have lost their lives to COVID-19,
and to their families and loved ones.

To the 250 million patients
who were infected, fought, and survived.

To the more than 46,000 patients who
participated in the vaccine's clinical trials.

To my Pfizer colleagues and their families
who made the impossible possible.

To our partners at BioNTech, and in particular
Ugur Şahin and Özlem Türeci.

To my wife, Myriam, and my children, Mois and Selise,
who stood by me and supported me throughout.

Contents

Foreword

IT HAS BEEN THIRTY-FIVE YEARS since my own moonshot, when I, along with my Carter Center colleagues, set out to eradicate Guinea worm disease, a horrific parasite that afflicted more than 3.5 million people primarily in impoverished parts of Africa. Today, there are fewer than thirty cases of Guinea worm disease in the world.

This goal led to the Carter Center's decades-long work to improve global health by fighting six preventable neglected tropical diseases in tens of thousands of overlooked communities that lack healthcare resources. Unlike COVID-19, which impacted nearly every living person on the planet, these diseases are mostly forgotten, therefore overlooked by the developed world.

The largest lesson learned from tackling these diseases is the importance of partnership and collaboration across sectors to drive global public health for all, and most important, to never give up.

Our work would not be possible without the innovative pharmaceutical industry. Medicine is central to eliminating and eradicating many diseases. In fact, a program of which I am most proud is the Carter Center's work to eliminate blinding trachoma in partnership with Pfizer Inc. and other organizations for more than twenty years.

We are making real progress, with eleven countries validated by the World Health Organization as trachoma-free in 2021. My hope is that we can achieve the global elimination of trachoma before the year 2030. Our collaboration with Pfizer has been fundamental to the fight against trachoma, and I am pleased to introduce the story of Pfizer's COVID-19 vaccine development. It was a moonshot the world needed and is a testament to Dr. Bourla and Pfizer's deeply held conviction that "science will win."

Pfizer was the first company to receive US Food and Drug Administration approval for Emergency Use Authorization for a COVID-19 vaccine. Dr. Bourla's purpose-driven leadership inspired the men and women at Pfizer to make the impossible possible. They deserve our appreciation for their resolve to take on this disease. I have been fortunate enough to visit Pfizer and have seen firsthand their pride in their work. I was grateful, but not surprised, when they took on this task—building hope when the world lacked it, believing in the power of science when the world doubted it, and challenging themselves to not give up.

You will marvel at the intricacies of Pfizer's moonshot: what it took to discover, develop, and bring to market the first COVID-19 vaccine.

—President Jimmy Carter
Plains, Georgia

Preface

Luck Never Comes to the Unprepared

"Excellence is never an accident.
It represents the wise choice of many alternatives—
choice not chance, determines your destiny."
—Aristotle, 384–322 BC

ON TUESDAY, DECEMBER 31, 2019, Chinese authorities alerted the World Health Organization to a mysterious virus causing pneumonia-like illness in a small cluster of patients in the city of Wuhan. Shortly after, the novel virus was identified as SARS-CoV-2. Less than a year later, on Tuesday, December 8, 2020, nearly ninety-one-year-old Margaret Keenan received a Pfizer/BioNTech COVID-19 vaccine shot at England's Coventry University Hospital and became the first person in the world to be vaccinated against the most devastating pandemic of the last one hundred years using messenger RNA (mRNA) technology. Wearing a cheerful Christmas sweater, she received a standing ovation by nurses and hospital workers as she was rolled through the hallway in a wheelchair. At the same time, all over the United Kingdom, people were celebrating like it was the end of a war, not the launch of a new vaccine.

The story of the nine-month moonshot that led to this glorious day takes place in those anxious times of 2020. But the story has its origins at least two and a half years earlier. On January 1, 2018, I was named Pfizer's Chief Operating Officer, a position that would give me one year to prepare for an eventual transition to the top. My focus was growth, and my motto was that "growth never just happens—it is created." And in our industry, the only way to drive growth is by creating a meaningful, positive impact on the lives of patients. For that we had to transform Pfizer into a patient-centric organization focusing on science and innovation. I am an optimist, perhaps because my mother's brave and narrow escape from death, minutes after being lined up against a wall in front of a Nazi firing squad during the Holocaust, made me believe that nothing is impossible. But my optimism for the new Pfizer I was about to lead was grounded on a strong foundation. During the past nine years, my predecessor Ian Read, a man of strong convictions from whom I learned a lot, had turned around our research and development

(R&D) engine from an organization of mediocre productivity to that of an industry leader. That gave me confidence to think "really big" about what to do to transform Pfizer and move "really fast" with executing my plans. In the twelve months that followed, before I became Chief Executive Officer, I developed the strategies, designed the organization, and decided on the executive team that would join me in our transformational journey. On the day I was named CEO, I knew exactly what I wanted to do. That day, the Pfizer board of directors called me into a conference room in a California hotel to confirm my appointment. I thanked them, smiled, and said aloud, "Only in America!" Only in America could a Greek immigrant with a thick accent become CEO of one of the world's biggest corporations.

Immediately after taking over the reins, I initiated the most dramatic transformation in the history of our company. Within months, I reorganized the portfolio of our corporate businesses. We found better homes outside Pfizer for our Consumer Healthcare and Upjohn (off-patent) businesses. Those two businesses were big. In 2018, they accounted for over 25 percent of our total revenue, but the first had low market share and the second was declining. In both cases, the new homes were offering stronger and brighter prospects for these businesses than if they were to remain inside Pfizer. Our Consumer Healthcare business was merged with GlaxoSmithKline's Consumer Healthcare business in a joint venture that created the largest and best over-the-counter company in the world. Our Upjohn business was merged with Mylan to form Viatris, the largest and best specialty generic drug company in the world. The separation was not easy. Some members of my team became concerned with letting that much revenue go. "We will not be the biggest anymore," someone said. "We should not aim to be the biggest. We should aim to be the best," I replied. "A good gardener needs to prune the tree when spring starts," I told a reporter. "Pfizer is in the spring of high growth." Little did we know at the time that by letting these two predictable but slow-growing businesses go, and focusing solely on our innovative core, we'd position ourselves two years later to deliver a vaccine to curb a global pandemic while reemerging as the biggest pharmaceutical company in the world. In addition to the worries about "size," there were a lot of emotional concerns. Some of the most iconic brands that made Pfizer famous were part of those two businesses: Advil, Centrum, Lipitor, Norvasc, Viagra, just to mention a few. It felt that we were separating ourselves from our biggest successes. But I knew that what makes a good company great is its ability to optimize its own successes and turn the page to new and better horizons of fame.

At the same time that we were reducing the size of the company, we invested billions of dollars to strengthen our scientific capabilities and pipeline assets. Within months, we acquired four biotech companies. Among them was Array, a Colorado biotech firm with a reputation for making "druggable the undruggable." These companies came with more expenses than revenues but were helping us build a position of scientific strength. Additionally, we started building new capabilities. For example, the first day I started as CEO, I appointed Lidia Fonseca as Pfizer's first Chief Digital Officer, reporting to me. Lidia, who was born in Mexico, emigrated to the United States as a child, and received her master's degrees in the Netherlands, has a passion for how digital solutions can improve health outcomes. She joined us from Quest Diagnostics and was an experienced change agent. One of the first priorities I gave her was to digitize research and development, allowing for greater collaboration, transparency, and speed. But all these moves were costing a lot of money, and to support this new direction, we had to do a radical reallocation of capital. The R&D and digital budgets were increased drastically. To offset this, we took draconian measures to reduce marketing and administrative expenses. Six months after my appointment as CEO, Pfizer had been transformed from a conglomerate of businesses to a company singularly focused on scientific innovation.

I knew that a business transformation of that magnitude could not be completed by just changing the portfolio of businesses or the allocation of capital. To be successful, we would have to change the company's culture. We would need to become a company more comfortable with risk and driven to achieve bold moves to live up to our purpose and its promise to patients.

With an aging global population that is becoming increasingly urbanized, it became clear to me that the demand for new breakthrough medical solutions will continue to grow. We needed a culture that could foster innovation by thinking big and out of the box. With the cost of healthcare delivery also growing, healthcare disparities could become an even bigger issue. We needed a culture of greater sensitivity to the needs of society and a dedication to a higher purpose. But a new culture is not something you can create in a vacuum. You cannot ask a consultant of a business school to suggest to you the best business culture so you can choose. The winning culture for an organization depends on its legacy, the institutional memory of its failures and successes, the challenges and opportunities in the decade ahead, the environment, and many other factors. You need to assess all these pieces of the puzzle and design your own winning culture.

Just two weeks after I became CEO, the top one thousand Pfizer leaders from around the world gathered in Florida to discuss, debate, and determine what a new Pfizer would look like. When we left, there was no doubt about our purpose. We exist because society needs—requires—us to deliver "Breakthroughs that change patients' lives." We had debated a lot if we should use the words "medicines" or "vaccines," but we decided to go with "breakthroughs," because it was broader and more powerful. With the lines between technologies becoming blurred, speaking only about medicines or vaccines seemed very restrictive and no longer reflected the realities of the new decade of scientific inquiry.

Just a few months later, we also launched an effort to introduce Pfizer's new culture to our employees and the world—in four simple words.

COURAGE: Breakthroughs start by challenging convention—especially in the face of uncertainty or adversity. This happens when we think big, speak up, and are decisive.

EXCELLENCE: We can only change patients' lives when we perform at our best together. This happens when we focus on what matters, agree on who does what, and measure outcomes.

EQUITY: Every person deserves to be seen, heard, and cared for. This happens when we are inclusive, act with integrity, and reduce health-care disparities.

JOY: We give ourselves to our work, and it also gives to us. We find joy when we take pride, recognize one another, and have fun.

Of course, no one could imagine in January 2019, when we spoke about "Breakthroughs that change patients' lives," that within two years our company's change would, in turn, change so much the lives of so many. We would develop the world's first safe and effective vaccine to be approved in multiple countries against the global pandemic. We succeeded not because we were lucky. We succeeded because we were prepared. The people of Pfizer learned in those preceding years that having the courage to think big and make tough, counterintuitive decisions is not only allowed but expected; that excellence in execution is not about us but about serving patients; that equity and reduction of healthcare disparities are not the concern of the others but an important part of our purpose, and that the good we bring to the world should spark our joy, pride, and passionate dedication to our purpose. Courage, Excellence, Equity, Joy. These powerful words became part of our lives at Pfizer and prepared us for the challenges ahead.

The idea of a moonshot is resurgent today. The term's first known use was in 1949, when Americans contemplated space exploration. Coincidentally, this was also a time of great progress in vaccine development with the combination of diphtheria, tetanus, and pertussis in one DTP vaccine. A polio vaccine was in use a few years later, by 1955. But the idea of a moonshot entered the lexicon for good in the 1960s when President Kennedy pledged that man would fly to the moon and return safely to Earth. Kennedy said he chose the moon, not because it was easy, but because it was hard: "Because that goal will serve to organize and measure the best of our energies and skills, because that challenge is one that we are willing to accept, one we are unwilling to postpone, and one which we intend to win."

More recently, Mariana Mazzucato, professor in the economics of innovation and public value at University College London, in her book *Mission Economy: A Moonshot Guide to Changing Capitalism*, writes that we saw many "spillovers" from Kennedy's moonshot affecting life on Earth—technological and organizational innovations that could never have been predicted at the beginning. It was a "massive exercise in problem-solving." These are the reasons I felt it appropriate to name this book *Moonshot*. Like Kennedy's moonshot, the work to develop our novel vaccine against COVID-19 was indeed a massive exercise in problem-solving, an exercise that allowed us to consolidate scientific knowledge of a decade within nine months and that will have spillover effects in many other scientific areas, affecting life on Earth more than we thought at the beginning.

The nine months you'll read about in this book were the most challenging and the most rewarding of my life, both personally and as a leader. The story of our success is the story of three colliding attributes—the power of science, the importance to society of a vibrant private sector, and the enormous potential of human ingenuity.

Today, all of us find ourselves facing enormous challenges, ranging from climate change to social division, inequities, and a plethora of problems in every community. I am sharing the story of our moonshot—the challenges we faced, the lessons we learned, and the core values that allowed us to make it happen—in hopes that it might inspire and inform your own moonshot, whatever that may be.

—Dr. Albert Bourla

Moonshot

1

Business NOT as Usual

"What matters is not what happens to you,
but how you react to it."

—Epictetus, AD 50–135

I WAS IN MY FIFTEENTH month as the company's CEO. The novel coronavirus that had begun as a worrying epidemic in Wuhan, China, was fast becoming a terrifying global pandemic. Our team in China had already been forced to work from home. The first known death in the US had been reported near Seattle, and suddenly cities, sports leagues, and the stock market were in disarray. President Trump sent me an urgent invitation to join other leading pharmaceutical and public health scientists for a meeting to be held on March 2 in the White House Cabinet Room. I was in Europe to deliver a keynote speech at the Delphi Economic Forum, and I had asked our head of research and development, Mikael Dolsten, to attend. A physician and scientist, Mikael was trained in Sweden and has participated in the creation of more than thirty approved drugs. He joined Pfizer in 2009 as the Chief Scientific Officer and head of research and development. We've now worked closely together for years. He would play a critical role in our moonshot, despite the personal challenges COVID-19 caused him.

The meeting was to take place Monday, and my phone rang Sunday around noon. Mikael had arrived in Washington, DC, a day early and was wrestling with the message we wanted to land with the Trump administration. Much of our deliberations up to that point had focused on developing therapeutics that would help keep patients alive. But what about a vaccine that would prevent people from getting the virus in the first place? In other words, would we point our resources toward a treatment or a prevention?

In the context of COVID-19, a treatment alone wouldn't end the pandemic. A vaccine could. A June 2018 paper from Vaccines Europe cites vaccination as "one of the most cost-effective public health interventions ever implemented" and makes the point that childhood vaccination "is one of the greatest medical success stories of the 20th century." According to the World Health Organization, vaccines prevent 2 to 3 million deaths each year worldwide. Yet every year, around

1.5 million people die from vaccine-preventable diseases around the world, 42,000 in the US alone. Globally, one in five children under five does not have access to lifesaving immunizations. Apart from clean water and sanitation, vaccines have had the most profound impact on public health. This was the reason why when I became CEO I significantly increased the investments in vaccine research.

On that call, Mikael and I agreed that this virus was different, and we warmed to the idea of pursuing a vaccine first. One aspect that distinguishes Pfizer's vaccine-making capabilities is that we are highly integrated, end to end, from early research to late-stage trials and clinical development. We also had one of the world's premier vaccinology teams, led by a tough, courageous German scientist, Kathrin Jansen, and as a result we had an ace up our sleeve. With a career that spanned breakthroughs at Wyeth, Merck, and GlaxoSmithKline, Kathrin came from an East German family that fled just before the building of the Berlin Wall. Trained as a microbiologist in Germany, she did her postdoctoral research at Cornell University, and today she leads a team of globally recognized scientists who work in Pfizer's laboratories in a small New York town called Pearl River. The director of the Vaccine Education Center at the Children's Hospital of Philadelphia, Paul Offit, told a reporter at *Stat*: "She's exactly who you want in that position. She fights for the vaccines she thinks are important. People who think pharmaceutical companies are evil should spend time with people like Kathrin Jansen."

I remember telling Mikael, "If not us, then who? Tell the White House that we are all in for a vaccine." I could hear the excitement in his voice.

The next day in the White House Cabinet Room, the president was flanked by Vice President Mike Pence and Health and Human Services Secretary Alex Azar, the National Institutes of Health's Dr. Tony Fauci, leaders from the Centers for Disease Control (CDC), and our colleagues from the pharmaceutical industry. There was a lot of discussion about treatments. When it came to his turn to speak, Mikael told the president, "It's not just one solution. I think we should offer multiple approaches, therapeutics and vaccines." He said that Pfizer was already advancing a therapeutic option and dedicating all resources necessary to pursue a vaccine at the same time, with a thirty-thousand-person team across the company to design clinical studies and manufacture both therapeutics and vaccines. He stated that Pfizer had started mobilizing our vaccine experts at sites around the world to prepare for this second approach against COVID-19 and that we would share with everyone around that table what we were learning, act as one team, and move swiftly given the urgency of the situation. "That's fantastic," the president replied. "Thank you very much. That's, really, very exciting."

Mikael called me later that day to report on the meeting. He told me that most of the discussion was about treatments, not prevention. "Tony Fauci seemed to be getting excited when I said we will go for a vaccine as well," Mikael said.

In the meantime, the world started to change as the virus spread. The forum I had been invited to speak at, the Delphi Economic Forum, was suddenly and urgently postponed. For me this triggered a major alarm, and I decided to return home early. On the plane, I kept thinking about the emerging situation with this coronavirus and what the Pfizer priorities should be in response to it.

Coronaviruses are a large family of viruses named for the crown-like protein spikes on their outer surface. This was not the first time we'd had to worry about them. In 2003, the outbreak of severe acute respiratory syndrome (SARS) caused worldwide concern, with its swift global spread resulting in thousands of infections, and approximately 10 percent mortality. Laboratories in Hong Kong and Germany, and the US CDC, confirmed the root cause to be a novel coronavirus. The disease was controlled within months. A few years later, in 2012, another novel coronavirus, which may have transferred to humans from infected dromedary camels, became the cause of another respiratory syndrome, the Middle East respiratory syndrome (MERS-CoV). However, now the situation with COVID-19 was looking different. For example, in China we had to shut down our facilities to protect the safety of our people from COVID-19. This was not something we'd had to do with SARS or MERS. Would we have to do the same with our facilities around the world? I also knew from China that in the first weeks of COVID-19 the hospitals in the affected areas were overwhelmed. Could that happen in other parts of the world? And if yes, would these hospitals have enough medical supplies to treat the patients who would storm their intensive care units (ICUs)? For us at Pfizer, this question was translated into, Would we be able to supply them with enough quantities of the medicines they need? I started imagining dark scenarios where the global demand for hospital medicines was growing exponentially and at the same time our manufacturing sites were shutting down because of the disease and were unable to supply enough treatments.

Finally, I kept asking two questions: What if the medicines that physicians had currently in their hands were not effective enough against this virus, as seemed to be the case in China? And could Pfizer do something about it? I knew that in 2012 we had tested some molecules against MERS that seemed to have good antiviral activity. Maybe we could test them against this virus as well.

It was a long flight to come back home, and I ruminated for hours and wrote thoughts in a Pfizer board of directors notebook that I had pulled from my bag. Before landing, I cleared my notes by consolidating a few bullets and deleting those that didn't seem important enough for the moment. I took a new piece of paper and I wrote on it what I thought would be the top three priorities for Pfizer:

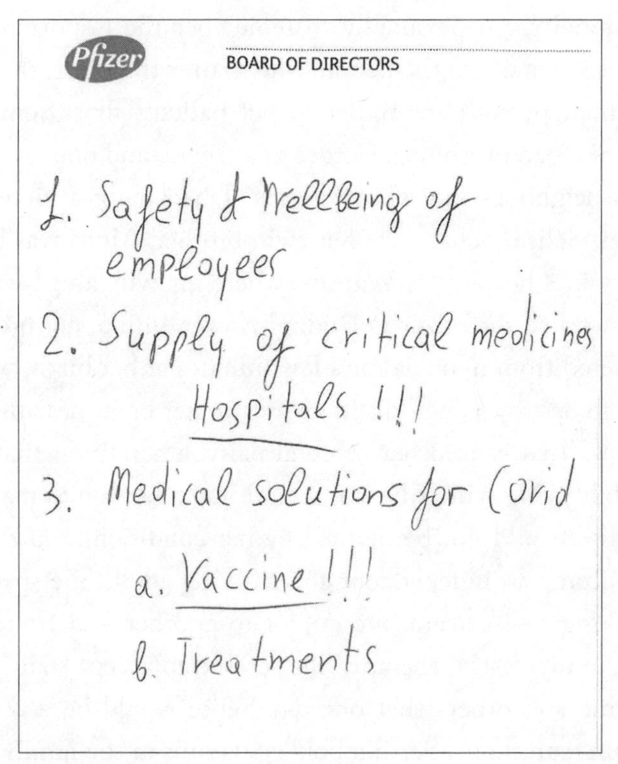
Pfizer
BOARD OF DIRECTORS

1. Safety & Wellbeing of employees
2. Supply of critical medicines Hospitals !!!
3. Medical solutions for Covid
   a. Vaccine !!!
   b. Treatments

Albert Bourla's handwritten note, written on the plane returning home from Greece as he contemplated the company's priorities in the early days of the COVID-19 pandemic, March 2020
Image courtesy of Pfizer Inc.

The next day I opened the door to our executive leadership team (ELT) meeting room at Pfizer's headquarters near the United Nations building in New York City. We had named this room the "Purpose Circle." When I'd taken over, I had removed the conference room table so that we could all sit together in comfortable chairs arranged in a circle. At first not everyone took to this arrangement, preferring the traditional setup. One element that warmed even the naysayers to the concept was a wall that prominently featured our purpose, "Breakthroughs that change patients' lives." On the other wall, each of our executives had hung photographs of patients who had inspired them personally to remember the importance of the decisions we make to the people who matter the most, the patients. It was their personal reminder to put patients first. Some hung a picture of a parent, some a picture of a friend, and one had a picture of their neighbor's kid who was sick. I had hung a picture of my daughter, Selise. Selise, like her twin brother, Mois, was born prematurely in a hospital in Warsaw, where my wife and I were living while I worked for Pfizer Poland. However, unlike her brother, Selise suffered from hypoxia for a few minutes at her birth, which was enough to destroy some cells in an area of her brain near the cerebral ventricles. That caused her cerebral palsy, a condition that affected her mobility. My wife, Myriam, had devoted her life to making sure that Selise would not be defined by her condition, and that Selise could become an independent and thriving adult. She spent all her time taking Selise from one doctor to another and from physical therapy to myofascial therapy. She refused to accept "fate" and kept telling me and others that one day Selise would be a dancer and could run marathons. Setting bold goals runs in the family. Myriam spent hours every day searching the Internet for new medicines under development, like cell therapies, that, if successful, could help Selise rebuild some of the destroyed cells in the area of her brain that was affected by the few minutes of hypoxia. Thanks to Myriam's sacrifices, Selise today is a bright student at Barnard College and lives an independent and full life in the college dorms. Her degree plan (at the time of this writing) is urban studies because she cares a lot about building more equitable cities. Her aim is to serve others in ways that will help people become happier and healthier. Unfortunately, those new medicines didn't come in time to help Selise rebuild her cells and dance or run, or even walk without a cane. But I knew that one day they could come in time to help other kids and us at Pfizer to do something about it, so this day might come sooner than later. Selise's picture on the wall of our executive team meeting room was my reminder to always put patients first. That day, in that room, many important decisions related to the three priorities I wrote on the plane had to be made.

But originally COVID-19 was not supposed to be the topic of discussion that day. We were supposed to continue our discussions for our ongoing major reorganization efforts. Not surprisingly, the decision we'd made in 2019 to increase our investments in R&D and digital capabilities, coupled with the need to reduce our cost base in marketing and administration, had created a lot of anxiety and tension. COVID-19 was not yet the central focus it would become. Leaders were still figuring out where they stood in the re-org. A CEO must have a spider sense for tension within the leadership team. I was sensing in early March 2020 that our effort to reimagine Pfizer's organizational structure was causing emotions to boil over. To let off some steam, I planned a dinner with my executive team for that night at the aptly named Break Bar on Ninth Avenue in New York. "The Break Bar is everything you'd expect from any other bar with a BIG TWIST (OR SMASH). Here, when you are done with your shot, pint or wine . . . you smash the glass" is how it advertises itself. You can book a Rage Room at varying levels of intensity—"Anger Problems" or "RAMPAGE," for example. I felt it was a creative way to let my team know that I understood how they felt. The dinner never happened, though. The COVID-19 lockdown canceled all plans. At our weekly executive leadership team roundtable in mid-March, the day we were supposed to go to the Break Bar, COVID-19 was the main concern in the air. I made it clear that it was time to rise to the occasion. I discussed with them the priorities I had set on my flight back: The safety and well-being of employees. Maintaining supply of critical medicines. And developing new medical solutions against COVID-19. We started making decisions.

One of the first and most difficult at the time was to close all our office facilities around the world and work remotely. We debated this question before the city and state of New York issued lockdown orders. When I asked to go around the room to hear from each person, it was clear there were differing views. After hearing them all, I had to make a decision. I decided to do it. "We have to close all offices and start working remotely." I turned to Lidia Fonseca, our Chief Digital Officer, and told her, "You must make this work. There is a lot at stake here."

For me this decision didn't come easy. I am an extrovert. I gain energy by interacting with people. It's important to look someone in the eye, read body language, walk down the hall with someone after a difficult conversation. How could I work from home? I was surprised to find, once we started working remotely, how easy the change came for me. Walk through the kitchen, turn left and you find the laundry, turn right and you find my home office. It is a relatively small room, suitable for no more than four people. But the size would no longer matter. In the next eighteen months, I could fit dozens of people on my new computer screen. I could still look them in the eye and read their body language, at least from the parts you could see on the screen.

Once this decision was made, someone asked, "What will happen to the people that are working in our buildings but are not our own employees?" As is typical with many large organizations, many people working in the mailroom, security, cafeteria services, and sanitation are employees of other companies that have contracts with us to provide their services. We work with these people daily, and everyone thinks about them like members of the family. In fact, one of my first thoughts was, What would happen to Luis? Luis Perdomo is a barista in our cafe, and he is a bit of a legend at Pfizer. He is also the epitome of joy. His smile lights up the room, and for many of us who drink coffee, he is the first person we see when we come to the office. Though his life has not always been easy—he tragically lost his teenage son to cancer a few years ago—you can always count on Luis for a smile, a kind word, and, if you're really lucky, a morning hug. He is a true gem, so much so that we forgive him for his undying love of the Boston Red Sox!

The concern we were all feeling was that if we closed the offices and discontinued some of these services for a prolonged period, Luis and many other wonderful service providers would likely lose their jobs. "Let's maintain regular payments to all these contractors under the condition that they keep the people working for us regularly employed, even though they won't provide their services to us," I said. Everyone felt good about it. Our procurement team took the responsibility to make this happen, working with each of the businesses within the company to keep contractors paid wherever possible. These early decisions took care of the first priority I had written on my flight back home, "Safety and well-being of employees." I was happy because, like other CEOs, I felt tremendous responsibility for the health and welfare of the ninety thousand Pfizer employees around the world. But I also felt we had additional responsibilities as a healthcare company in the middle of a health crisis. Our people discover, develop, and produce medicines for hundreds of millions of patients globally, people who rely on us each and every day—people suffering with cancer, heart disease, or arthritis. Our research centers and manufacturing plants around the world would have to remain open. My immediate fear was expressed in the second priority I had set for us, "Supply of critical medicines." COVID-19 or not, people had to receive their medicines. I was concerned with potential shortages of needed medicines—stockouts, as we call the situation. I was particularly concerned with hospitals. They were expected to get overcrowded, and Pfizer is one of the largest suppliers of injectable hospital medicines in the world. I discussed this concern with Frank

D'Amelio, our long-standing and highly regarded Chief Financial Officer, who likes to roll up his sleeves and problem-solve. Frank, who also has the responsibility of overseeing manufacturing, warned me that stockouts would be inevitable. The demand for these medicines could be ten to fifty times higher, and at the same time it would be difficult to ramp up production under COVID-19 restrictions. I rallied our team, imploring that we must deliver in the middle of a health crisis. Failure was not an option. They acknowledged the challenge and immediately jumped to a wartime footing. We already had in place a crisis management plan that was developed years ago with situations like this in mind. The management of manufacturing had already activated this plan, and our manufacturing sites were operating under "preparedness level 2," which involved certain restrictions and special safety measures. We decided to elevate all sites to "preparedness level 3." Only people essential for the operations would be allowed in, and much stricter safety guidelines would be implemented. In the next few weeks, I was amazed to see the dedication of our manufacturing workers. The number of absentees remained at less than 3 percent. At the same time, I felt an obligation to all those workers who had to report every day to the manufacturing sites while the rest of us were working from the safety of our homes. I felt I needed to make a gesture of support, and I asked to visit one of our manufacturing sites to express to the workers my solidarity. My request was denied by their management. My chief of staff at the time, Deb Mangone, my trusted thought partner of many years, called to give me the news.

"Albert, they said you should not go."

"Why?" I replied.

"Because you are not considered essential," she said, and laughed.

This was the moment I realized how professional these people are and felt confident that we were on the right track.

Returning to the Purpose Circle, the time we had allotted to this meeting was approaching its end, but we still had to discuss our most impactful priority: new medical solutions against COVID-19. Immediately after the sequencing of the virus was released, we started looking in our molecular library for antiviral compounds that matched. We identified several that were promising, but more tests were needed. Additionally, only a couple of days earlier, Mikael and I had agreed to commit to the president of the United States that we would work on a COVID-19 vaccine. We discussed both: the potential to develop antiviral treatments and a new vaccine that could work against COVID-19. But these programs cost a lot of money. Money that was not budgeted. Frank D'Amelio reminded us of this and asked how much money we were talking about. I still remember Frank's facial expression when Mikael gave him his ballpark projections.

"Ouch, Mikael!" Frank said as he wrote the number in his notebook.

To make things worse, Angela Hwang, President of our Biopharmaceuticals Group, expressed concerns that lockdowns would significantly and negatively affect our forecasted revenue streams of our current business. Angela leads all our commercial activities and is one of the savviest business leaders I know. So I took very seriously her concerns that we were facing the worst of both worlds for a publicly listed company. Not only were we about to spend way more than we had budgeted, but we were also going to have lower revenue than we had projected. As we were all struggling with the reality of the numbers, I asked everyone's opinion on this issue. They all felt the same: we must do it. In situations like this, with so many lives at stake, financials should fall lower on the priority list. I was proud of our team, all of them, and I pointed toward the wall with the pictures of our beloved patients and said:

"Clearly this is not business as usual. If we miss our budget for a year, no one will remember it the year after. If we miss the opportunity to do something for the world now, we will all remember it forever."

At the end of the meeting I went through the room to each of my executive leadership team members and assigned them specific tasks related to what we had agreed. I sensed the team's resolve. And they sensed in me that we were going into battle together. It was the last time in 2020 that we met in the Purpose Circle, but the groundwork it laid prepared us for the battle ahead. This day was the beginning of our moonshot.

2

What Is Obvious Is Not Always Right

"You will never do anything in this world without **courage.** It is the greatest quality of the mind next to honor."

—Aristotle, 384–322 BC

DURING THOSE HARROWING NINE MONTHS in 2020, we had to make hundreds of difficult decisions. Many of them came down to me. The pressure was as high as you can imagine. It's not an easy thing when you feel the hopes of billions of people and millions of businesses and hundreds of governments invested in this industry, and you are the leading company in this industry, and it happens that you are the CEO. And also a *new* CEO. You feel that pressure on your shoulders. And you feel that you need to rise to the occasion. What is at stake is beyond imagination. It is the health of the whole planet in a pandemic that we haven't experienced in a hundred years. It is the fate of the global economy as countries around the world become paralyzed by lockdowns and practical fears. And then, to make it even worse, it is also a political debate, in the middle of an American presidential election, in a time of deep polarization. It shouldn't come down to politics, and yet that is a pressure too. I needed to navigate all that and still try to lead my team to deliver to the world a solution that would save lives.

Our effort was not without shortcomings, but we were fortunate to get more decisions right than wrong. Most important, we were fortunate to get the most critical ones right. And as usual in life, the most critical decisions were the most challenging to make. When I look back, the one that clearly stood out was the decision to use the mRNA technology to develop a COVID-19 vaccine. Not only because a different choice would have yielded very different outcomes. It was also because that particular decision was the most counter-intuitive. The obvious option was *not* to use mRNA. It required a lot of forward thinking and eventually a lot of courage, but this is what gave us the vaccine.

When I asked our team to develop an effective vaccine in record time, the options they could choose from were abundant. For Moderna, for example, the critical question would be, Should we try to develop a COVID-19 vaccine or not? Which technology to use would never be a question for them. They were highly specialized in mRNA, and when they decided to pursue a vaccine this would be for them the obvious and the only option. But for us, it was different. Our research team had experience with many available technological platforms upon which to build the vaccine, including adenovirus, recombinant proteins, conjugation, mRNA, and others. My first challenge to them was to recommend which platform we should bet on. They debated among themselves, and to my surprise they came back to me with the suggestion to use the mRNA platform. Two years earlier we had partnered with a German biotech company called BioNTech, founded in 2008 by a kind, charismatic, and humble husband-and-wife team to focus on cancer treatment. Their mRNA technology, we felt, had the potential to help us with our efforts to create a much more effective seasonal flu vaccine. I was a big fan of the mRNA technology and certainly believed we had a good chance to transform the development of flu vaccines, but I was thinking that we were still a few years away from this advancement. Mikael broke the news to me—we'd be pursuing an mRNA solution—and my first reaction was surprise. Selecting this technology to develop a COVID-19 vaccine was not the obvious choice.

While mRNA vaccines are not a magic bullet for every epidemic or pandemic of the future, in this pandemic they literally helped save the world. So before I describe my discussion with Mikael, the mRNA vaccine story merits being fully told.

In nature, mRNA is a single-stranded chemical molecule that is complementary to one of the two DNA strands of a gene. The DNA holds all the information required for the body to form and function. For example, the instructions on how to produce a protein—let's say a hormone that is critical for our function—are always coded in our DNA and are transmitted genetically from generation to generation.

Dr. Ugur Şahin and Dr. Özlem Türeci, the husband-and-wife cofounders of Pfizer's COVID-19 vaccine partner, BioNTech
(Credit: BioNTech SE/Stefan Albrecht)

When our body needs to produce this hormone, it will copy these instructions on an mRNA molecule and send it to an organelle called a ribosome, which will move along the mRNA, read its instructions, and produce the hormone.

There are many different types of vaccines, used to help prevent infections, all of which have the same goal: to train your immune system to recognize and defend against infectious disease agents, which are called pathogens. Vaccines usually contain weakened, dead, or noninfectious parts of these pathogens. They cannot cause the disease, but your immune system will recognize them as a foreign invader and will summon its defenses against them, i.e., antibodies and T-cells. When the real pathogens come along, the body will be ready with millions of antibodies and T-cells that will attack them immediately and with force, reducing their chances of causing disease.

The novel mRNA vaccines, however, are different. They are not made up of an actual pathogen. They don't contain weakened, dead, or noninfectious parts of a virus or bacterium. They contain instructions on how our body could produce proteins that are part of these pathogens' construct. When our ribosomes receive the instructions by reading the mRNA that we injected, they start producing these parts of the pathogen, which our immune system will recognize immediately as invaders. Our system will then mobilize an immune response that will protect us from the real pathogens if they come along. In a nutshell, mRNA teaches the body to make its own vaccine. The injection is a set of instructions for how to protect yourself.

Pfizer's Vice President and Chief Scientific Officer of Viral Vaccines, Philip Dormitzer, had joined us in 2015 from Novartis after they decided to exit their vaccine business. Phil led a team that developed a synthetic approach to influenza vaccine strain updates and pandemic response. They realized that synthetic technology could potentially be used to improve flu vaccines by matching more closely the circulating strains and could be delivered in greater quantities and more rapidly than vaccines produced by conventional technologies. In 2012, Phil and others published an article showing that self-amplifying RNA encapsulated within a lipid nanoparticle potently elicited antibody and T-cell responses. These findings were early signals of what was to come. Pfizer has been interested in RNA in part because of its ability to respond quickly to changes and be consistent. Although molecules of RNA also can vary in their behavior, one piece of RNA behaves a lot more like another piece of RNA in general than either viruses or proteins do. We liked the flexibility of the technology compared with traditional vaccine technologies. This flexibility includes the ability to alter the RNA sequence in the vaccine to potentially address new strains of the virus, if one were to emerge that is not well covered by the current vaccine.

In 2018, Pfizer had turned to Kathrin and her research team to identify a partnership that would advance mRNA development for a game-changing seasonal flu vaccine. During the partnership search, Kathrin made fast friends with the Turkish-German cofounder and CEO of BioNTech, Ugur Şahin. Ugur came to our vaccine R&D center in Pearl River, New York, and presented BioNTech's approach to mRNA. He later told me that at the beginning Kathrin had a lot of pointed questions for them, but eventually he could see that she was convinced. We recognized almost immediately that, unlike other firms we'd spoken with, BioNTech was agnostic about the type of RNA that might work. In other words, they refused to favor one approach over the other. They came from the mercurial and mysterious world of oncology, so they were curious and open to different approaches. They were scientifically sophisticated, but they also recognized the role of intuition. The teams at Pfizer and BioNTech clicked immediately because of similar worldviews, and a three-year research collaboration agreement was signed. During this period, BioNTech would transfer to us technical expertise and licenses that would allow us to develop a novel influenza vaccine. I was the COO of the company at that time, and when the agreement reached my desk for approval, I executed it immediately. The size of this contract was relatively small by Pfizer's standards, so I didn't ask to see the CEO of the company as I would typically do with larger deals. Little did I know then that our mRNA influenza vaccine work would give us a profound head start when COVID-19 struck two years later.

In January 2020, as COVID-19 spread in China, our partners at BioNTech were among the first to jump on the virus, studying its sequencing and effects from data on the Internet. Chinese researchers published the virus's genetic sequence on January 11, 2020. The virus was spreading in a way no one could understand, much less contain. A major vaccine effort was needed, but BioNTech would need a partner. Ugur thought of us, and given the positive interactions and trust established during flu collaboration, he called Kathrin. But before Kathrin received the call, I had already asked her and her team to suggest the technology platform we should use for our own vaccine efforts. She had decided to suggest mRNA. And Mikael Dolsten FaceTimed me to let me know the suggestion. I was surprised.

"Mikael, to be honest, I didn't expect that," I told him. "It would be a very risky and complicated bet."

My doubts were grounded on facts and rational thinking. First, the technology was promising but not proven. If we were successful, this would be not only the first COVID-19 vaccine but also the first ever mRNA vaccine that the world would see. In contrast, the adenovirus and protein technology platforms, with which we had even better familiarity, had delivered multiple other successful vaccines in the past. Second, we would have to negotiate an agreement with BioNTech, a process that typically takes months to complete. We wanted to move fast, and negotiating contracts didn't feel easy to do under time pressure. Third, BioNTech was small, and we likely would have to absorb all the development and manufacturing costs and realize, in case of failure, significant losses alone, but we would have to split the profits with them in case of success. I pointed out all these concerns and discussed them with Mikael.

"Are you sure about this?" I asked.

He was. Mikael was convinced through our work with flu that the technology was the right choice.

"The technology is ideal for something like this. It is fast, and capable of being edited quickly for updates and boosters. Adenovirus or other viral vectors technologies could have difficulties with boosters because the immune system will create antibodies not only against the coronavirus but also against the adenovirus," he said.

Mikael knew from previous discussions that the speed of development and the ability to boost frequently were very high priorities for me. With a forethought that was sadly correct, I was afraid that by the winter of 2020 we would have a new and probably more lethal wave of COVID-19, as happened with the 1918 influenza pandemic a century ago. We needed to have a vaccine by then. I also knew that a virus with these characteristics would mutate sooner rather than later. It was important to have a vaccine that we could boost as often as was needed without the fear of losing potency. I started seeing his rationale but continued the discussion.

"What about using protein technology?" I asked.

"We are good with proteins, and we could certainly make a vaccine with this platform, but with mRNA we would have both humoral and cellular immune responses. With proteins, we will have good antibodies, but I am not sure we would have T-cells," he replied.

We discussed a little bit about potential manufacturing hurdles, how a development program with mRNA could look, and at the end I expressed my concerns about having to partner with another party in a hurry. Mikael felt that Kathrin Jansen had developed good relations with the founders of BioNTech and signing a contract would not be a challenge. "Okay, Mikael," I said. "Let's gather the team to discuss their proposal." Mikael was relieved and pleased that I didn't reject the idea up front and that I was willing to explore it further.

A few days later, Mikael assembled Kathrin and her team of scientists. In addition to Phil, Kathrin had invited Bill Gruber, our Senior Vice President of Vaccine Clinical Research and Development. A board-certified pediatrician and pediatric infectious disease expert by training, Bill had decades-long experience with respiratory diseases. He obtained his undergraduate education at Rice University and attended Baylor College of Medicine for medical school, residency, and a postdoctoral fellowship in pediatric infectious diseases. He then joined the faculty at Vanderbilt University, where he focused his research work on viral respiratory infections, notably respiratory syncytial virus (RSV) and influenza. RSV is a common respiratory virus that usually causes cold-like symptoms. Prior to COVID-19, we were on the fast track to develop an RSV vaccine, and we had accumulated already significant knowledge about what works and what does not. The relevant part for the discussion was that RSV, in addition to being a respiratory disease, also has a similar spike like COVID-19. Other very talented scientists were also on the videoconference.

"How would you like to do this, Albert?" Kathrin asked.

"Let's start by focusing the discussion on the opportunities. What are the benefits we get by using this technology?" I said. "Then we can discuss the challenges and what we can do to overcome them."

During that meeting we discussed in great detail the things that we had conceptually discussed with Mikael two days ago. They explained that this technology would enable our researchers to design and modify the mRNA very rapidly. Because mRNA is synthetic, it doesn't contain live viral particles, which makes it a much more well-defined product. It would allow our team to administer the vaccine several times if needed—not just during the pandemic but also later if an immunity boost was needed because of virus mutations. Where a traditional type of vaccine might take months to design, this mRNA version would need just a couple of weeks. They told me that they had very good expertise with this technology and that they felt the technology was mature enough to give a vaccine. When we discussed the challenges, Phil spoke up.

"You know, with mRNA we can move quickly, but the vaccine will have to be frozen to remain stable during distribution."

I hadn't thought about it before, and I realized that this could be a major disadvantage.

"What can we do about it?" I asked.

"We can develop later a more stable formulation, but at the beginning we will have to deal with it. Manufacturing should find a way to manage it."

Then I asked if we thought BioNTech would agree to work with us on this. Kathrin said that Ugur had already called her to express an interest. I felt that a decision had to be made. Clearly, selecting the mRNA platform was a much riskier and more complicated option than all the others we had available. But it was the fastest way to a solution, and my entire team was 100 percent behind this choice. Vaccinologists represent a special character in life sciences. All physicians and life scientists are called to the field of medicine to help people, but I've found that people who get involved with vaccines are different. First, they tend to come to the industry to work only on vaccines, and they are extremely committed to protecting people. Second, they are generally more conservative. Unlike medicines, which are designed to treat disease and chronic conditions, vaccines are about prevention. Therefore, this approach engenders a completely different mindset. Vaccinologists must weigh risk and benefit differently than most physicians. If I am treating a patient with cancer, for example, I calculate risk differently than if I am protecting a healthy individual. Vaccinologists face a lot of scrutiny in terms of the safety and efficacy of their research programs and as a result are very conservative with their choices. Knowing that, I realized how deeply convinced they must be to suggest such a risky option. And they were the best experts in the world. My gut told me that this was the right choice, and I told them.

"Okay, mRNA it is. I will call their CEO tomorrow."

I had never met Ugur in person before. We had never spoken even on the phone. During the previous two years that we were collaborating on a flu vaccine there was never a need for us to speak. He was always talking to Kathrin or Mikael. I took the initiative to call him, introduce myself, and express my deep dedication to the success of this program. We started talking, and from the first moments we hit it off. The personal chemistry was there and was instant although our personalities were profoundly different. I am an extroverted Greek Jew who immigrated to the US. Ugur is an introverted Turkish Muslim who immigrated to Germany. I felt immediately that I could trust him. We discussed the principles of our collaboration. We would let science drive our decisions. Above everything else, the goal would be to deliver the safest and most effective vaccine we could. Each one of our companies would focus its efforts on the areas of its specific expertise, but we would be fifty-fifty partners on everything. At the end I told him that time was of the essence, and I asked him if he would feel comfortable starting the work before we signed a contract.

"Ugur, it could take months to finalize all the needed agreements, the research, manufacturing, and commercial contracts."

"Albert, your word is enough for me," Ugur said. "We can start the research work immediately. The lawyers should focus on the research agreement first, and when it's ready we can sign it. The other contracts can wait."

I immediately agreed with that. I felt that with BioNTech, we could and would have to think differently in a crisis that demanded a novel approach.

The next day, our team and the team at BioNTech in Germany held a wide-ranging conference call to discuss candidate antigens, toxicology study plans, first in-human clinical trial plans, regulatory interactions, and manufacturing schedules. They discussed disaster planning if one or more sites were to go down for whatever reason. They agreed on the basic plans and started having daily meetings. A couple of weeks later, with plans already well advanced, we signed a letter of intent so that the lawyers from both sides could be reassured that at least we had something in writing. On April 9, we signed a collaboration agreement to codevelop a first-in-class, mRNA-based coronavirus vaccine, aimed at preventing COVID-19 infection. BioNTech received from us a $72 million up-front payment and was eligible for future milestone payments of an additional $563 million for a total consideration of about $636 million. In addition, we were going to provide $113 million in cash to them by buying equity from the company (bringing our total ownership of the company to about 2.3 percent at that time). Under the collaboration agreement, the two parties agreed to share all development costs and profits fifty-fifty, but Pfizer agreed to cover all these costs up front. If the project failed, Pfizer would bear all the losses alone. If the project succeeded, BioNTech would pay back to Pfizer its share of development costs from its profits from the commercialization of the product. Although we didn't agree to all the terms two companies normally would address when starting a project of this size and complexity, we did agree that BioNTech would have commercialization rights of the potential vaccine in Germany and Turkey, and Pfizer would hold these rights in the rest of the world. China was excluded from this agreement, as BioNTech already had an agreement with a Chinese company for China, Hong Kong, Macau, and Taiwan. The collaboration agreement also called out specific agreements that the two parties would need to negotiate and conclude relating to the collaboration agreement. These were the manufacturing and commercial agreements.

In the months that followed, we dedicated our entire focus to our efforts to develop, get approval for, and manufacture this vaccine. We never found the time to conclude the manufacturing and commercial agreements in 2020. We continued cooperating under the initial letter of intent, the collaboration agreement, and, just as important, our mutual trust, until January 2021, when we signed the commercial agreement.

3

Thinking Big Makes the Impossible Possible

"Our problem is not that we aim too high and miss,
but we aim too low and hit."

—Aristotle, 384–322 BC

UNDER NORMAL CIRCUMSTANCES, THE DEVELOPMENT of a vaccine takes years. Many of these projects fail. In the case of HIV, scientists have been working on a vaccine for decades, and we still don't have one available. From discovery to development to approval to manufacturing to distribution, new vaccines must adhere to a highly regulated process before reaching people's arms. The process requires patience and perseverance. Scientists in the lab work for years on an idea rooted in basic science before they are allowed to develop several "prototypes" for further testing. In a highly reiterative process that usually also takes years, these prototypes are tested in tubes and animals and then sent back to the lab for adjustments that will better the idea and theoretically improve the perceived efficacy or safety. After these adjustments, they are retested again and again in a process that aims to discover the best candidates. We do preclinical and toxicological studies, reactogenicity studies, virus neutralization studies, immunogenicity studies, efficacy studies, and many other types of highly specialized studies. In most cases, the tests and assays that are necessary to assess and evaluate the different prototypes do not exist, and we will have to create those too before we can test the prototypes. Every time we see something that we do not like—for example, fever in mice or weak virus neutralization—we will run additional experiments to understand these effects and suggest changes to the prototype. Of course, the molecular engineers who are working on these prototypes confront many other worries as well. For example, the prototypes will have to be chemically or biologically scalable. This means that, if successful, they can be manufactured with high quality at large scale; otherwise, the development of a promising prototype could come to nothing. This is not a given all the time. On many occasions we have stopped developing promising molecules because it was impossible to reproduce them outside the lab at a large scale.

If these preclinical studies in the lab are successful, they will eventually identify a few candidates that give us good reason to believe they will be safe and effective. Then we begin our clinical studies in humans. First, we start with studies that will allow us to identify the right dose. These are Phase 1 studies on healthy volunteers (subjects) that we call dose escalation studies. We always start with a very small dose and put the subjects under close medical supervision. We are doing tests and looking for any signals that could create any safety concerns. If everything is approved to go forward, we move to a higher dose and repeat the same process. At the same time, we measure the biological effect that each dose has on humans. We cannot estimate efficacy at this stage; instead we are looking for surrogate endpoints, possible measures of effect, like stimulation of the immune system with antibody and T-cell responses. We progressively increase the dose until we reach a level that is safe and satisfies our surrogate criteria for efficacy. If this is not possible, we go back to the molecular scientists to reengineer the candidates, and once again we will start new Phase 1 studies with new vaccine candidates. We usually do this several times until we are either successful or decide to discontinue the program.

If our Phase 1 studies are successful, we move the best candidates into Phase 2 studies. In the Phase 2 studies, we try the vaccine candidates in different regimens—one or two doses, three or six weeks apart, in young or old adults, etc. We are trying many combinations, and we constantly measure safety and efficacy surrogate endpoints until we decide that we have one that is optimal. A Phase 2 study usually has very high standards for determining success because it green-lights the initiation of the final and most important study: Phase 3. The learnings from these earlier phases inform the refined focus necessary to initiate a Phase 3 study. This phase demands the most significant resources and manpower, with the intent of generating the data necessary to receive regulatory approval. If it is successful, you have a vaccine. If it fails, the program is usually terminated. Therefore, it is imperative to give it your best shot by selecting the best candidate and the best regimens in the Phase 1 and Phase 2 studies.

The Phase 3 studies are crucial and required for regulatory approval. These studies follow very strict rules and standards that have been established by regulators, like the Food and Drug Administration (FDA) or European Medicines Agency (EMA), to ensure integrity and scientific rigor. Typically, they run for a long period of time and involve thousands of volunteers (referred to as participants), monitored by hundreds of independent physicians (called investigators), working in independent institutions (usually hospitals) called investigation sites. Typically, in these studies we are comparing a test group to a control group. The participants in the test group will receive the vaccine under evaluation. The participants in the control group will receive either another vaccine that is already registered and represents the current standard of care (if a vaccine against the same disease already exists) or a placebo (if there is no other vaccine for this disease yet). Depending on the study design, the participants and the investigators don't know who has received the placebo and who has received the vaccine. The two are presented in identical vials with barcodes that only a computer algorithm can identify. This algorithm also ensures that the two participant groups have a similar composition in terms of gender, age, health status, etc. This way we can be certain that the results will reflect the differences between treatments only and will not be affected by differences in the composition of the two groups. Usually, in the Phase 3 trials we do not base our evaluation on surrogate endpoints. We measure real efficacy. The participants, after receiving their shots, continue living their lives as before, but their health is monitored. Some of them will be exposed to and infected by the disease we are investigating. When this happens, the investigators monitoring them will confirm with lab tests that the subjects have indeed contracted the disease we are studying and will record that information. Of course, at that moment they don't know if these patients had received the vaccine or the placebo. As part of the clinical trial design, the statisticians in the study protocol and the regulators, an independent group of experts that form the data monitoring committee, will predetermine the number of total confirmed positive cases that trial participants have to reach in order for the next step to occur. Once we have reached this number, the data monitoring committee initiates the process of unblinding the data so they can see how many of the diagnosed cases belong to the test group and how many to the control group. This committee will then inform us of their decision: we should either *continue the study* (when the difference in positive case numbers between the two groups has not reached statistical significance), *stop the study for futility* (when the data indicates that statistical significance between the test and control groups cannot be achieved), or *stop the study for efficacy* (when efficacy has already been achieved with statistical significance and there are no safety issues).

When both the efficacy and the safety of a vaccine are established through a Phase 3 study, it triggers the next two steps. The regulatory group will start preparing dossiers with all the data required by regulators for approval. Typical vaccine dossiers for initial applications will run thousands of pages, and each one of them needs to be quality controlled for accuracy, a process that is also time-consuming. At the same time, the manufacturing group starts the preparations for the industrial production of the new vaccine. This will require ordering raw materials and in some cases specialized equipment that usually costs hundreds of millions of dollars.

I was keenly aware of all these complexities when I asked our research team to bring me a plan for an effective and safe vaccine with timelines that would look nothing like the ones we'd seen before.

As COVID-19 infections and deaths rose in early 2020, the world faced an unprecedented crisis, and we had to rise to the moment. We had a fantastic vaccine team staffed with some of the most brilliant scientific minds in the world. They immediately started the work. A few weeks later, in an April video call meeting, Kathrin Jansen and her team presented an aggressive plan that could bring the results of a pivotal Phase 3 study to completion by the second half of 2021. Their plan depended on condensing years of required work into eighteen months. At the same meeting, Mike McDermott, the head of our manufacturing group, presented plans that would allow us to develop a manufacturing process within eighteen months, including finding appropriate suppliers of raw materials and designing from scratch new specialized equipment that would probably be required for mRNA manufacturing. "I can have tens of millions of doses just months after we know we have a vaccine," Mike said. Mike directed all of this from his home office, which was decorated with drawings made by his five daughters. His plans would smash any previous record of drug development speed and manufacturing scale-up.

The team was exhausted and proud to have come up with such an aggressive plan. But in the meantime, the pandemic had taken a dramatic turn for the worse, particularly in New York City, where many of us were living and working. Hospitals were overflowing, ICUs did not have enough respirators to treat patients, and bodies were stacked in refrigerated trucks outside because all the morgues were full. Every night when I went to sleep and every morning when I awoke, I felt the anguish, the toll this disease was taking. Infection rates and deaths grew higher and higher. With entire economies at risk, the quest for an answer went beyond the future of any one organization. It was about the future of the world. Since making the decision with Mikael to go for developing a vaccine, I had come to feel more and more that this would be the only effective solution.

And we didn't have time. I knew that a century ago, during the 1918 influenza pandemic, the second wave was much more lethal than the first. I also knew that during the coming fall we would face the double danger of a flu season amid the pandemic.

"It is not good enough," I told the teams. "We must have it by this October. And we must have *hundreds* of millions of doses by next year, not tens of millions."

I still remember their surprised faces projected on my computer screen. It was not a look of disappointment that I was rejecting their plan. It was shock and confusion over what exactly I was asking them to do. They felt it was impossible. We debated for a while why this could not be done from their perspective. Their facts were solid, and their arguments clear. Rationally, they were right. It could not be done. But it *had* to be done. There is a time for consensus-building, and there is a time to push. This was the time to push. I told them that the plan was not accepted—period. I asked them to go back and rethink everything from the beginning and make the impossible possible. They should have zero concerns on costs. They should have no consideration for return on investment. They could have all the resources they needed to get it done. They should think about doing things in parallel instead of sequentially. They should design their experiments in clever, innovative ways that would allow us to learn fast and eliminate vaccine prototypes rapidly so that we could decide on the final one quickly. They should build manufacturing capacity at risk before knowing if we had a product or not. They should procure all needed raw materials that were readily available and place committal orders to purchase those that were not. At the end, I asked them also to include a final slide with calculations of how many people would die in October if we failed to produce such a plan. A week later, the team came back to me with a genius plan that, if successful, would bring us results by the end of October 2020.

The design of the Phase 1 and 2 programs was done in a very clever way. Instead of waiting to have all candidates released from the lab before they started the program, they would start the study immediately, when the first vaccine candidate became available. They would test it on multiple combinations at different doses, in different regimens, and with different age groups so they could develop a good understanding of how the candidate interacted with the immune system. Once the second candidate became available, they would follow a smart plan of targeted tests, comparing it with the first one. These tests would allow us to draw conclusions for the second candidate without repeating the entire set of tests we had done for the first. We would repeat the same with the third and fourth candidates. The idea was to quickly terminate the less promising candidates, concentrate on the best two, and with a few additional tests select the final candidate that would move to a Phase 3 study. At this stage of their presentation, they warned me that although this design would give them useful information to inform the selection of the best candidate, the risk of making a mistake and moving to Phase 3 with a suboptimal candidate was great. In hindsight, this decision turned out to be one of the most critical, in a period where it felt like we were making life-and-death choices every day.

"And as you will realize in a moment, this Phase 3 study will cost more than whatever you have seen so far," Kathrin Jansen told me.

I took a mental note as they continued their presentation.

The Phase 3 study was designed to give conclusive results in the fastest possible way. It was going to be a placebo-controlled, double-blind study, meaning that neither the participants nor the researchers would know which treatment they received until the study was over and unblinded. It would be randomized one to one, meaning an equal number of participants would receive the final vaccine candidate and the placebo. The FDA had set a limit of at least 50 percent efficacy for a vaccine to be approved. Our team had designed the study to be able to demonstrate vaccine efficacy of 60 percent, an even higher internal standard. The FDA usually asks for two months of safety data for Emergency Use Authorization and six months of safety data for full approval. Regardless, our team would follow up with the enrolled participants for two years. The statistical analysis that our mathematicians performed indicated that for this level of efficacy (60 percent), they would need a total of at least 164 events of COVID-19 (participants who get the disease) to demonstrate statistically significant efficacy. A study size of ten thousand to fifteen thousand participants could potentially generate these results in less than a year (dependent on incidence rate of disease/infection). They decided to blow out the study size and go for thirty thousand participants so they could accumulate COVID-19 events faster (later we increased this number to over forty-six thousand). A study of this size usually involves 40 to 60 research sites. They decided to target 120 sites so they could recruit these participants faster (later we increased this number to 153). But the most critical factor was to select research sites in areas where there would be a heavy COVID-19 disease burden, so the attack rates (percentage of participants who contract the disease naturally during the study period) would be higher. If you have low attack rates, and as a result fewer people getting sick, you cannot be sure if your vaccine provides protection. We needed high disease intensity for the clinical study to demonstrate that vaccinated people were better protected than the placebo recipients. The problem was that the attack rates for different locations were changing over time. When a city or county had a lot of infections, they would usually take policy measures, and the attack rates would drop after a period of time. On the contrary, when a city or county did not have many infections, people would start to relax and the COVID-19 infection rates there would climb higher over time. How would our researchers know where to place our research sites so they would have access to high attack rates within the time that was needed (at least seven days after the second dose)? Our epidemiologists developed an algorithm that could predict as accurately as possible the areas where COVID-19 attack rates could rise at certain periods of time. The research team would then try to open investigation sites accordingly, so that by the time the participants had received their second doses, the chances of high disease burden would be greater. As the team continued articulating their plans, I was thinking, *Wow*.

After the vaccine team completed their presentation, Mike McDermott, the head of manufacturing, took the floor. While the lab would prepare the vaccine candidates to be tested in Phase 1 and Phase 2 studies, the manufacturing team would start working to scale up the manufacturing process even without knowing which candidate would be eventually selected. They would need to order raw materials for all candidates to be ready for all scenarios, and they would eventually have to dispose of those that were specific to the candidates not selected. Our challenge was that there was no industrial production, anywhere in the world, of any mRNA product (medicine or vaccine). Until now we had only produced mRNA in laboratories at small scale for research needs. Our manufacturing team would have to invent, design, and order new industrial formulation equipment that didn't currently exist. Mike confirmed that our engineers had already started the design work for the formulation equipment and had begun discussions with specialized equipment producers to understand what it would take to build these machines fast.

"What about the challenge with the ultracold storage requirements?" I asked. "How would you store millions of doses at negative seventy degrees Celsius?"

"Thanks for asking. For storage we have a good solution. We will move materials for our existing product out of our current warehouses to temporary locations and convert these warehouses to freezer farms, each the size of a football field."

"Freezer farms?" I asked.

"Yes, freezer farms. We will install five hundred big freezers that each has the capacity to store three hundred thousand doses. In total we can store over one hundred million doses there. We will do the same in our manufacturing site in Belgium for the European production."

I thought he could drop the microphone there, but he continued.

"Of course, you will now ask me how we will ship, under these ultracold requirements, millions of doses to thousands of locations around the world. This is a big challenge indeed. As you know, there are no cars or airplanes that can hold these low temperatures. So sending them the same way we send other vaccines or medicines that require cold chain won't work. By thinking out of the box, we realized that the solution was inside a box. What if we had a container that is relatively low-cost and could maintain temperatures of negative seventy degrees Celsius for let's say one to two weeks. We can do that by filling them with dry ice."

My heart was filling with joy, but I interrupted him to ask what I thought would be a challenging question.

"Where are you going to find that much dry ice?"

"We calculated that we would need one to two percent of the US dry ice supply. The problem is more logistical, but we have a solution. We will start manufacturing dry ice in our facilities and use it on-site."

I thought that was bold and asked him to continue.

"We could send these boxes with normal vehicles, trains, or airplanes to any location in the world. Upon arrival, they could either store them in special freezers that are commercially available or keep them in the box for longer by replenishing the dry ice. I asked our engineers to start the design of a box like this that in fact can be reusable. We would also have in the box an electronic device that has GPS, thermometer, and light detector. The device would transmit in real time the location and the temperature to our control center. If someone opens the box, the light detector will also transmit this information to us. Albert, we would know everything about each box that is traveling around the world until it reaches its destination safely."

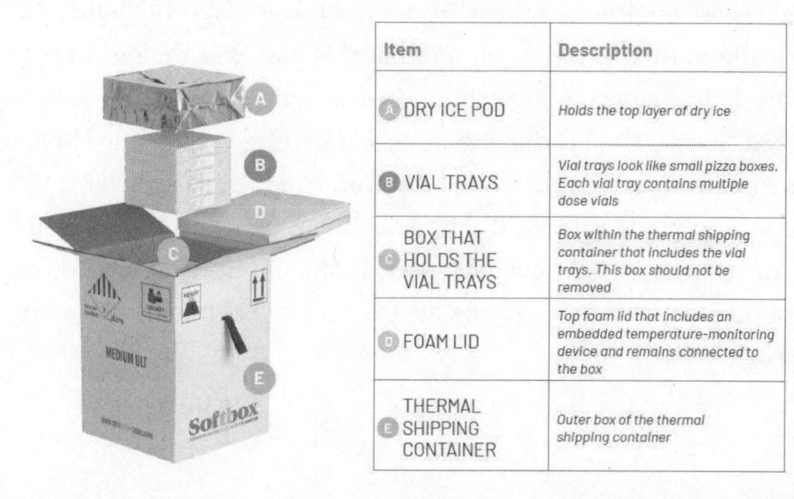

Item	Description
A DRY ICE POD	Holds the top layer of dry ice
B VIAL TRAYS	Vial trays look like small pizza boxes. Each vial tray contains multiple dose vials
C BOX THAT HOLDS THE VIAL TRAYS	Box within the thermal shipping container that includes the vial trays. This box should not be removed
D FOAM LID	Top foam lid that includes an embedded temperature-monitoring device and remains connected to the box
E THERMAL SHIPPING CONTAINER	Outer box of the thermal shipping container

Breakdown of a Pfizer COVID-19 pallet shipper, showing the critical components used to track the shipper's temperature controls
Image courtesy of Pfizer Inc.

The temperature gauge on Pfizer's COVID-19 shipper, created to ensure that the company could track the temperature and location of each pallet of vaccines while en route to patients
Photo courtesy of Pfizer Inc.

The plan could cost as much as $2 billion. I congratulated the team, thanked them, and gave them the go-ahead. I took a note to inform our board of directors of this decision. It was a very expensive bet, and I knew that it would be painful to take a $2 billion write-off in my second year as CEO if the project failed. But I also knew that it would not take the company down, and it was the right thing to do. I called our lead director, Shantanu Narayen, to discuss it with him. Shantanu is the chairman and CEO of Adobe, and since I became CEO he has been a mentor to me. He is the voice of reason, and he has a unique ability to bring people together, which he does often with the other directors of Pfizer's board. He listened carefully, and he agreed with me that this plan was the right thing to do. I then called a few other board members to make sure they agreed. It was obvious that everyone was getting very worried about the spread of the disease around the world and felt that Pfizer could and should play a role. And the criticality of the situation called for an all-in approach. A few days later, I announced to the world our intention to discover a vaccine against the global pandemic by the end of October.

4

Lightspeed

"The fight doesn't wait for those who delay."

—Aeschylus, 525–456 BC

ON MARCH 19, 2020, WE held a meeting focused exclusively on the vaccine. On my calendar, it read, simply, "COVID-19 Vaccine Plan." After that, the session became a SWAT team meeting that to this day we call "Project Lightspeed," and it appears as such on my daily calendar. It was apparently a coincidence that a few months later the Trump administration used the name "Operation Warp Speed" for their COVID-19 response task force. Our project was named "Lightspeed" to make expectations clear to all participants. Everyone should adopt the energy of light speed, 186,000 miles per second, to progress their work. But even if everyone was doing so, still that would not be enough. A project team has its own dynamics, which causes events to move at the speed of the team rather than the speed of the individuals. The work of one is dependent on the work of the others. When committed individuals come together, the force and effect of their collective efforts accelerate even more—that is, the sum is greater than the parts. And, in our case, fast decision-making by a dedicated and motivated team would be the key determinant of success for this project.

Usually, in the corporate world, you need multiple disciplines to be consulted before any decision can be made. But the biggest challenge for every corporation is to reach a decision among peers who disagree. This was the reason why in 2016, when I was the head of our biopharmaceutical business (Pfizer Innovative Health), I had broken the division into six business units, each of which had relative autonomy to make tactical decisions within their budget. Vaccines was one of them, in addition to Oncology, Internal Medicine, Rare Disease, Inflammation & Immunology, and Consumer Healthcare. I told the global presidents of these business units that they should each consider their unit like an entrepreneurial biotech company, with the presidents of each business unit acting as CEOs and with me in the role of the CEO of a private equity firm that owns them.

"A private equity firm does three things with the biotechs they own," I told them.

"First, it appoints their management. I have appointed you.

"Second, it agrees with their management on the strategic direction. We have discussed and have clear alignment on the strategic direction that each one of you should follow.

"Third, it allocates capital to them. You will have to compete for that. The best proposals will get the funding."

Every month I would review in a committee their requests for investments in research projects, manufacturing infrastructure, and commercial expenditures, and I would allocate capital to their approved projects. Then I would leave them relatively alone to execute their plans while I monitored metrics and outcomes.

But even after we implemented this system, most of a business unit's decisions were still quite complex and required broader consultation in a matrix of multiple stakeholders. Decisions still depended on painstakingly slow diplomacy and multiple compromises. Otherwise, things could stop. It might feel counterintuitive, but this is even more so when high-level executives are involved in a project. When a decision made by a multidisciplinary team at a lower level reaches the next management layer, it is a common theme that one of the "bosses" will have a different opinion, causing things to pause. When eventually the disagreement is resolved, the decision goes to the next management layer, and very often the scenario is repeated.

The new approach I had implemented had improved our agility and the speed of our decision-making, but for a project like Lightspeed we faced greater levels of complexity than ever before but still needed to reach decisions even faster than ever. For example, we had to consult many different scientists from different research groups, engineers and other supply leaders from different specialized manufacturing groups, lawyers, commercial colleagues, financial people, communications people, and the list goes on and on. All these people were vital to the decision-making process, as we were dealing with very complicated topics that required their very specific expertise. But getting all of them to agree before the team could move to the next step was imperative. We could not afford to let bureaucracy or egos slow us down. We needed to replace hierarchical processes, simplify the chain of command, and merge three to four layers of management into one fast-moving project team that could make decisions on the spot.

We held meetings twice weekly from 4:00 to 6:00 p.m., though often the meetings ran overtime. I led the meetings. Research, manufacturing, finance, legal, and corporate affairs were all represented with two, three, or more layers of management as needed. I acted as the "project manager." I confess that I am not the best project manager. I'm okay at it, but the value I brought was different. The CEO can knock down silos, hear everyone out, and quickly move things forward. Everyone on this team was encouraged to agree or disagree, oppose or encourage, no matter their position. But at the end of the meeting, decisions could be made fast because I was there and had the authority to make them. In typical meetings with the CEO, people usually have a lot of premeetings to "align" before the main meeting. With this project, they had very little time for that. Data flowed and decisions were made in real time. We had a just-in-time mentality. Not everyone felt at ease with this at the beginning. It can be uncomfortable to be together with your boss (or even the boss of your boss) and your own people. Pretty soon, though, the team's passion to cross the line as fast as possible with a vaccine before October 2020 replaced hesitations.

In the months that followed, every Monday and Thursday, from my home office computer, I saw a gallery of twenty-five or so leaders from across the company. Our agenda spanned basic science to clinical, manufacturing, and regulatory issues. Which vaccine construct should we select? Which dosing schedule should we follow in the trial protocol? Are we happy with how rigorous the testing is at a particular site? Why are we not recruiting patients for the clinical trials quickly enough? How do we ensure diversity in the clinical trials? Are we positioned to manufacture enough doses? How many doses per vial are being extracted? Let's review again the intricacies of the FDA's Emergency Use Authorization. Where are we on this provision? What is the latest with the Center for Biologics Evaluation and Research (CBER) at the FDA? Are we prepared for the Vaccines and Related Biological Products Advisory Committee (VRBPAC) at the FDA? The questions kept coming.

After a long day, Netflix provided distraction and a respite from endless meetings and preparations. I'd leave my home office, stop in the kitchen to get a glass of cold Chardonnay, walk into the family room, sit down on the couch, and watch an episode of *Game of Thrones* or a French spy thriller, *The Bureau*. The kids thought it was funny that my wife, Myriam, and I would binge on *Gilmore Girls*, a comedy-drama that reminded us that there were other problems in the world. Sometimes I would pause the show, reach for my phone, and FaceTime with a colleague to discuss an idea or resolve a problem. Or I would jot a note for the next day. Between shows I might call friends or sit and speak with Myriam to hear about her day. If ever I think that I am busy, I remember how hard she works to take care of all of us. Always with a smile and always with positivity. When I speak about her with our closest friends, I describe her as someone between Superwoman and Mother Teresa.

In meeting after meeting, I found myself curious but also impatient. I was getting fascinated by the details of this cutting-edge technology and was asking questions to understand it better. Sometimes I would ask these questions during the meeting, but many times I would call an expert afterward and ask him or her to give me details or explanations for something that was discussed in the meeting. I saved those questions for after the meeting so that I did not waste everybody's time. I also constantly questioned every step of the process and challenged every single timeline. If someone said it would take weeks, I asked why it couldn't be done in two days. This could be irritating to many at the beginning, but pretty soon it became second nature for the team to come proactively with solutions that would speed up our timeline, often attached to a hefty request for additional funding. But money was not the issue here. Time was. Very frequently when approving the additional investments, I would tell this team, "Time is life," repurposing the expression "Time is money." After a few meetings it became clear to all what was expected.

Looking back, I think this attitude—*Time is life*—was the most important success factor for this project. Setting goals that are very aspirational, goals that someone has never achieved before, can unleash human creativity in phenomenal ways. When you ask people to do in eight years something that normally takes ten, they will find it challenging, but they will think of solutions within the current process. If you ask them to make three hundred million doses instead of two hundred million (that was our current annual capacity at that time), they will find it hard but will investigate solutions that improve the current way of doing things. They may achieve something better by doing so, but usually these processes have been optimized over years and there is only so much you can do to deliver more. However, in this case I didn't ask people to do it in eight years. I asked them to do it in eight *months*. I didn't ask them to make three hundred million doses. I asked them to make three *billion* doses. And I insisted that these targets were not negotiable. It was clear from the beginning that incremental improvements would not make the cut. They needed to completely rethink their processes. They had to redesign them from scratch and be creative at every single new step of a new process. And they did it!

On April 22, 2020, we announced that German authorities had approved the start of our study to evaluate the four vaccine constructs against COVID-19, and the study began the following day. We had developed four vaccine candidates, each representing a unique mRNA format and target antigen combination. Two of the four vaccine candidates included a nucleoside-modified mRNA (modRNA), one included a uridine containing mRNA (uRNA), and the fourth vaccine candidate utilized self-amplifying mRNA (saRNA). Each mRNA format was combined with a lipid nanoparticle formulation. Two of the vaccine candidates encoded an optimized full-length spike protein, and the other two candidates encoded the receptor binding domain (RBD) of the spike protein. The RBD-based candidates contain a piece of the spike that was thought to be important for eliciting antibodies that could inactivate the virus, while the longer spike protein was thought to be important for eliciting a broader or more differentiated antibody response. The first studies in humans were meant to compare the four candidates at different dose levels. The dose escalation portion of the Phase 1/2 trial included approximately two hundred healthy subjects ages eighteen to fifty-five and targeted a dose range of one microgram (μg) to one hundred μg, aiming to determine the optimal dose for further studies as well as evaluate the safety and immunogenicity of the vaccine candidates. The studies also evaluated the effects of repeated immunization for three of the four vaccine candidates.

In late May 2020, we began to test these four different vaccines in Phase 1. Traditionally, those would have been tested sequentially, but we decided to test them in parallel—four different vaccine candidates, each at three different dosage levels. Such an effort might take a year under normal circumstances. We did it in a month. By late July 2020, we were prepared for a combined Phase 2 and 3 trial in which eventually over 46,000 patients would be recruited at 153 clinical sites in six countries.

As July 2020 approached—our target date for starting the pivotal Phase 3 efficacy phase—we faced yet another vexing decision. We had two promising final candidates, two different vaccine formulations. The first (known as b1) was the one that used only the RBD of the SARS-CoV-2 spike protein. This candidate had the most data and appeared to be an excellent choice. The second (known as b2) was the one that used the full-length spike protein. It appeared to have a broader immune response with fewer tolerability issues, fewer chills, and fewer headaches. Very preliminary data was also indicating that it might be potentially more potent in older adults, who were more susceptible to severe COVID-19 in the first place and whose immunity was more difficult to boost. However, b2 was more difficult to produce, and with less overall data available, it was a much riskier choice. Data for the first was further along because testing for that candidate had started earlier; data for the second lagged. With a key deadline approaching, we would have to decide which vaccine we moved forward with in Phase 3 testing. It was like having two puzzles—one with more pieces assembled that was revealing a beautiful picture and another with far fewer assembled pieces but revealing a picture with more vivid colors.

We huddled together as a team, studying the data we had and projecting what was possible. Even though there was a lot of white space in the second puzzle, it appeared to be more promising. With our original deadline approaching, we decided to wait another week for more data, because we sensed that the choice we made would make a huge difference in the success of the vaccine. Postponing our deadline by a week on this project was a very big decision. In fact, I don't recall making any other decision to delay a timeline of Project

Lightspeed by a week, not even by a day. For ten days we deliberated, fretted, and deliberated some more. Should we pick the one that we know is good enough, or do we reach for something that might work better? Eighty percent of the data we had was from b1, and we couldn't afford the time to develop additional data for b2. Every day delayed equaled the loss of many human lives. Some of the team felt that b1 would be good enough; others were not convinced. Still others were concerned that we were losing time going after something "perfect." "Perfect" is usually the enemy of "good," they would say. It was true that substantial work would be required of Pfizer to take the b2 construct and make it commercially viable—to manufacture it at scale with consistency.

In a critical meeting on July 24, 2020, arguments in favor and against each of the two candidates were summarized for the last time by the study team. The most recent data—only a few hours old—was also presented. It was time for a decision. In my computer monitor I could see all eyes looking at me. I had to make the final call. I made the riskier decision.

"You have done an excellent job articulating the pros and cons of both candidates," I told the team. "I feel that most of you believe that the best option is the second one. Let's go with the b2 candidate and let's hope we are right."

In the months to come, we held our breath that the Phase 3 efficacy and safety data would justify our decision, because once we started down this path, there was no turning back. It was an amazing relief when we finally could see the beautiful data set from the vaccine trial that had enrolled over 46,000 trial participants, and we were able to take that to the FDA and ask for Emergency Use Authorization.

The Phase 3 study started immediately. Ugur and I discussed the importance of diversity in our clinical trial at the start of our partnership, and we agreed during the spring that we needed to set the pace for racial representation. We worried about vaccine hesitancy among Black and Brown people around the world if their communities were not represented in the research. We welcomed the FDA's insistence that it be a priority for every company working on the vaccine. The vaccine had to be trusted. Historically, several minority groups have been underrepresented in research, including people of color and women. Black Americans account for roughly 13 percent of the US population but make up only 5 percent of clinical trial participants. Latinx people account for roughly 19 percent of the US population but make up only 1 percent of clinical trial participants. Pfizer has always done a very good job on this and was performing ahead of the industry standards. Pfizer's head of clinical development and operations, Marie-Pierre Hellio Le Graverand, a leader in Pfizer's clinical research, coauthored a paper in 2021 that examined the demographic diversity of participants in Pfizer-sponsored clinical trials in the United States. That study demonstrated that in contrast with the industry average, our overall trial participation of Black or African American individuals was slightly ahead of the US census level (14.3 percent vs. 13.4 percent), participation of Hispanic or Latino individuals was below US census (15.9 percent vs. 18.5 percent), and female participation was at US census (51.1 percent vs. 50.8 percent). The results provided a baseline upon which we could quantify the impact of our ongoing efforts to improve racial and ethnic diversity in clinical trials.

But the COVID-19 vaccine trial was not any average trial. It was the most important trial in the world at that time, and we had to get diversity right. Of course, we encouraged everyone to be represented. But the choice to participate in a clinical trial is a personal one and many times can be driven by bias or misinformation. We had to do more about it. A star in this effort was Sandy Amaro, who manages diversity in Pfizer's clinical trials. Her husband, Jean Amaro, also works at Pfizer in quality assurance. For Sandy, having married into a "very large and loving Dominican family" and raising two Dominican children, her work hits particularly close to home.

"I want to make sure my family is represented in clinical trials, in healthcare systems, and show our kids that hard work can change the world," she said.

Working alongside the entire team, Sandy's mission is to achieve statistically diverse participant pools for our clinical trials. Said differently, she is focused on equity and inclusion, supporting our clinical teams to have the right representation of participants. The goal of diversifying clinical trials is to better understand how a therapy or vaccine affects people of different ages, races, ethnicities, and genders, in order to detect any safety or efficacy differences in particular populations. In Sandy's own words, "We have to let the science drive us," and this is a philosophy we at Pfizer have built into the fiber of how we operate. In addition to following the data, Sandy and her team also recognize the importance of community engagement, working with multicultural advocacy partners, medical institutions, and legislative organizations to educate and encourage people who are at the highest risk or are historically underrepresented in clinical trials to participate. This is exactly what she did to ensure diversity in our COVID-19 vaccine Phase 3 trial. Through a very broad outreach, we recruited patients in African American communities in New Orleans and Atlanta, Native Americans in the Navajo Nation, and Hispanic/Latinx people in cities and rural areas nationwide. This was very important. The Navajo Nation, for example, would go on to outpace the US overall in the rate of vaccinations when the vaccine became available. In the end, 42 percent overall—and approximately 30 percent of US participants—came from diverse backgrounds, geographically, racially, and across age groups. These efforts and particularly these results didn't go unnoticed. In the spring of 2021, a leading patient research and consultancy group based in the UK, PatientView, upgraded Pfizer from the fourth-ranking patient-centric pharmaceutical company in the world to the second among fourteen in "big pharma."

For endless weeks and months, whenever someone raised a complication, a hurdle, or a challenge, I would respond, "People are dying. There is no excuse. Solve it." I knew this was a kind of psychological or emotional blackmail. But it was unfortunately true. Kathrin lived in New York City near a hospital and would tell us many times, as we were waiting for people to gather so we could start a meeting, that when she would go for walks during the lockdown, she'd notice the refrigerated makeshift morgues at what was then the epicenter of the pandemic. I would refer to her descriptions many times during our Lightspeed meetings. I truly felt bad reminding people about the stakes in such a graphic way, but I also couldn't resist because it was so effective. People were giving 200 percent of themselves because they knew that "time is life." And, of course, the pressure to keep timelines and deliver outcomes was relentless. Looking back, I never regretted that I pushed people hard. I know that without my doing so, we would have never succeeded, and the world would have been in a very difficult place today. *The Wall Street Journal* would later report on Pfizer's crazy deadlines and its "pushy CEO." But I truly regret that on several occasions I became unnecessarily unpleasant. I let the stress show rather than hide it. And knowing that I did that to people who were working day and night makes it even more regrettable. At that moment, most were feeling that I pushed too hard. But later, when they were accomplishing more than they thought they could, they all took such great pride in the impact they had on the world that they forgave me. But they didn't forget.

At Pfizer, all managers are rated every six months by their teams on how they live Pfizer's four values: Courage, Excellence, Equity and Joy. I had always performed at the top of the company in all these values. But in the survey results I received in late 2020, my Joy scores plummeted. It was not that I had been tough. People could accept that in the middle of a pandemic, even welcome it. But there were times I'd lost my temper in front of colleagues. I forgot to recognize the hard work of someone because I was too busy resolving the issues of another. Some cultural differences made this even more profound. In my Mediterranean culture, you use your voice for emphasis and meaning, and this can be very annoying for people from other cultural backgrounds. I can be an explosive personality. In fact, over the years I had to tone this down a lot in order to survive in the corporate environment of a global company, particularly when I arrived in the US. But now the weight on my shoulders of this big responsibility had betrayed me, and I'd given grief to many people who didn't deserve it.

I reflected on the feedback I received, and clearly the reviews were correct. I presented these results to my team and told them how much I regretted letting them down. Stressful situations test human character, but I should have known better. Humans can learn from their mistakes, and I became determined to not let this happen again.

In late October 2020, I sent an email to the eleven members of my executive leadership team.

> Team,
>
> Everyone I speak to—colleagues, elected officials, scientific leaders, investors, etc.—asks me when will we know if the vaccine works? I'm sure it's the same for you. Now the moment is near. In the coming days, the DMC [data monitoring committee] will likely give us an interim analysis on our COVID-19 trial. The world is watching. The stakes could not be higher.

I want to take this moment to tell you how proud I am of this leadership team. For more than 8 months now you've managed to run this global enterprise remotely without a hiccup. At the same time, you embraced the three goals we established back in March to: 1) care for our 90,000 employees, 2) maintain a steady supply of medicine to those who count on us and 3) bring a vaccine forward this year.

I know this team is focused on outcomes, not effort. Still, ahead of the readout, I want to say thank you for your extraordinary commitment. Win or lose on the vaccine (and I believe we will win), we've found breakthroughs in every area. Pfizer will never be the same.

On a personal note, I cannot wait until the pandemic is behind us and we can be together again in the Purpose Circle.

Albert

5

The Ultimate Joy

"The greatest pleasures come from the contemplation of noble works."

—Democritus, 460–370 BC

ON THURSDAY, NOVEMBER 5, PROJECT Lightspeed met for our regularly scheduled session. It had been two days since the contested presidential election. This meeting was to learn when we would have enough data to unblind the vaccine trial results and decide whether we could move forward on an Emergency Use Authorization application. It became clear that we were close. By Sunday night we would know.

In the hours leading up to the readout on the clinical trial data, a small group within the clinical development team quietly unblinded the trial data to tabulate effectiveness results and report to the data monitoring committee. That Sunday, November 8, would be the readout of data to our executive team, our first glimpse at what these nine months of research and development would mean. They set a time for Sunday afternoon. Behind the scenes, they scrambled to collate and tabulate reams of data from test sites around the world. They worked around the clock. Later, I would learn that one data analyst received his data packet at 1:30 a.m. but suddenly lost Wi-Fi connectivity at his house. He had to drive around town in the middle of the night looking for Wi-Fi in order to meet his deadline of handing off data to the next stage of analysis by 4:30 in the morning. He found a gas station, closed for the night, with a weak signal barely discernible outside the garage. As he crunched the numbers from his car, a police officer pulled into the station to ask him what he was doing. When the analyst explained to the officer he was working on the vaccine, the officer stayed with him to make sure he was safe. His data packet was transmitted in time.

On Saturday, November 7, the day before, I had to work hard to preoccupy myself. The sunny late-fall day did little to brighten my spirits. The tumultuous presidential election of 2020, which had politicized our vaccine efforts, had just ended a few days earlier . . . sort of. And, like everyone else around the country, I was on pins and needles. But for me, the anxiety for results extended beyond politics. The week was ending with news that more than 120,000 Americans had contracted COVID-19 in a single day, the highest number of new cases on any single day of the pandemic. Those numbers would continue to skyrocket. As the CEO of Pfizer, this news hit me as a personal affront. I felt tremendous responsibility.

Only a few of us inside Pfizer, alongside our partners at BioNTech in Germany, knew that the next day, Sunday, November 8, would be when we would learn the results of our Phase 3 trial of our mRNA-based vaccine. Dozens of questions were coming to my mind. Had we been right to select the mRNA technology? Was the decision to select candidate b2 instead of candidate b1 wise? Maybe we should have chosen boosting with a second dose twenty-eight days later, like everyone else, instead of twenty-one. And why had we chosen to test results seven days after the second dose instead of waiting fourteen days, when the immune response would be higher? Everyone else was testing at fourteen days. Were we brave or arrogant?

The next morning, I drove to one of our satellite offices in Connecticut, an hour northeast of New York City, to gather with a small team to learn the results. We had invested our people's passion, our science, and our technology in a gamble with enormous implications for our company and for humanity.

My car arrived at the office simultaneously with Mikael Dolsten's. When we met outside, we realized that although we had been working together every day, very intensely, for the past eight months, this was the first time we'd been physically together in months. We had made countless decisions together, and we had developed a breakthrough vaccine candidate, the results of which were just about to be read out, but in all this time, we had not met in person. I had met physically on occasion with other members of my team, but Mikael was very careful because of a personal situation at his home, so we did all of our meetings through Webex and FaceTime. For Mikael, the work we were doing to fight COVID-19 was truly personal. Of course, as a physician and a wonderful human being who cares deeply for patients, Mikael was devoted to the cause of developing both a vaccine and a treatment for COVID-19. Early in the pandemic, his wife, Katarina, who is also a physician, was infected by COVID-19 and hospitalized at Mount Sinai with serious symptoms. For many weeks, day and night, he cared for his wife, who was fighting the virus from the intensive care unit of the hospital. At the same time he was working on a vaccine and a treatment to ensure that no one would go through what his wife was having to go through. Myriam and I were good friends with Katarina, and we were seriously worried as well. I have known Mikael for many years, but I have never seen him more worried and disturbed than he was during the weeks his wife was battling COVID-19. I became reluctant to call him to discuss the Lightspeed program. He realized this, and he told me that his wife's situation motivated him even more to work on this project, and he remained deeply involved with it. As I said, for him, it was personal.

After we greeted each other with warmth and an elbow bump, Mikael and I went inside to meet up with the others. They were already there: Doug Lankler, our General Counsel and my close advisor, who manages an incredibly large number of complex legal and business challenges with wise judgment and good humor; Sally Susman, our Chief Corporate Affairs Officer who is charged with leading Pfizer's external engagements and whom I sometimes refer to as Pfizer's secretary of state; and Yolanda Lyle, a gifted attorney and my new chief of staff, who seamlessly stepped into the role during the pandemic. I sat down with these colleagues on whom I had come to rely for their confidence and calm, and enjoyed casual conversation about the weather, sports, and current affairs. Anything to make time move more quickly. Doug said he was so nervous he thought he was going to vomit. Anxieties aside, it was pleasant to be surrounded again by my friends and colleagues, but I was also mindful of the soft chimes coming from my chief of staff's mobile phone. Yolanda was to receive notification that the data monitoring committee had met, and that results were in. The stress, exhaustion, frustrations, hopes, and dreams of endless pandemic weeks were coming to a head. At 1:27 p.m., Yolanda's phone chimed.

"Assemble ELT, please."

The five of us quietly stepped into a conference room where we connected to a Webex video conference. Rod MacKenzie, our Chief Development Officer, was already connected from Michigan, where he lives. In the room there was also a documentary cameraman to capture the moment.

We all stared at the screen, but no one from the clinical trial team was there to tell us the news. We waited and waited. After a few long minutes, I joked that this torture was retribution for all of the pressure I'd put on this team over the long weeks and months of testing.

"It is payback day." I smiled.

To fill the time, we continued to chat. Mikael was sitting right next to me. I asked him to predict the level of efficacy we would hear from the trials. He squirmed in his chair and with some hesitation said 70 percent. I thought, *I hope he is right*.

Finally, the researchers with the findings connected. We tried to read body language, but they were not revealing much. There is that moment in movies about space exploration when mission control anxiously pauses for the crackle of sound from a distant astronaut calling from a capsule that has safely landed or passed to the other side of the moon. It was like that.

Bill Gruber, Pfizer's Senior Vice President of Vaccine Clinical Research and Development, said, "Good news. The study is successful. The committee of independent experts that reviewed the unblinded data strongly advises us to submit immediately a request for Emergency Use Authorization."

"Strongly" and "immediately" were unusually potent words for this committee. Their terminology over the course of these trials had always been careful, appropriately sparse, clinical. But now we could sense their enthusiasm. We jumped from our chairs and started celebrating. Sally, Doug, and Yolanda were screaming. I felt like I was wearing one of those wingsuits, flying above mountains and green valleys. Within moments, Yolanda appeared with a bottle of champagne that she had already chilled in anticipation of what she hoped would be positive results. As we toasted this amazing moment, I remember my eyes being drawn to the two Pfizer executive protection officers who were with us that day. They were silent as they usually are, but they had realized what was happening, and you could feel that they were seriously moved. One of them almost had tears in his eyes.

We continued celebrating, but we didn't know yet the best part of the story. It came fifteen minutes later when two experienced biostatisticians, one of whom was Satrajit Roychoudhury, reported the level of efficacy to Doug and me. We had agreed that only he and I would listen to the actual efficacy numbers so that we could decide how to proceed with this potentially material information. Everybody else left the room, and Doug and I connected to another video call to hear the results. I had thought that north of 60 percent effectiveness would be a good result. But one of the biostatisticians told us, "From the ninety-four cases of confirmed COVID-19, ninety belonged to the placebo group." I was shocked and felt I'd misheard, so I abruptly interjected.

"Did you say Nineteen? One–nine?"

"No. Ninety. Nine–zero!"

"But what is the efficacy?"

"Ninety-five-point-six percent, sir."

Doug and I remained speechless for a few moments.

"How conclusive is this number?" I asked.

"The statistical significance is very high, sir," one of them replied. "We do not expect this number to change much when we have all the one hundred sixty-four cases accumulated and proceed to the final readout."

We thanked the biostatisticians and exited the video. Doug and I looked at each other. I realized that we were probably sitting on the most material information in the world. The responsibility was very high.

"What do we do now?" I asked him.

"This information will have a significant impact on public health and will affect the way health authorities think and plan for the pandemic. We must disclose it immediately."

I nodded my head indicating my agreement. Previously, I had discussed with Moncef Slaoui, the head of Operation Warp Speed (OWS), that we would only disclose success or failure in the interim analysis and not provide a specific number before having final numbers. I recall Moncef assuring me that Moderna would do the same. Moncef's concern appeared to be that if we gave a specific efficacy number from the interim analysis and the final analysis was one to two points different, that could confuse the public. But we were all thinking between 50 percent and 70 percent efficacy. Now we were sitting on news that health officials around the world needed to know and prepare for. A vaccine with 95.6 percent efficacy was coming to the world. This completely changed the landscape. Doug, of course, was aware of what we had discussed.

"What about what we discussed with Moncef?" I asked.

"This is very important information for all health authorities in the world," Doug replied. "We must get it out there immediately."

I reflected for a moment and said, "What if we say we saw more than ninety percent efficacy? This will give the world a clear indication about the magnitude of the efficacy without disclosing a specific number, as we agreed with Moncef."

"That will work," Doug said.

We opened the door, and Doug went to the other meeting room to call Mikael, Sally, and Yolanda into our room. We closed the door, and the first thing I did was to remind Mikael of his previous prediction of 70 percent efficacy. And then I let the news fly.

"Mikael, it is more than ninety percent," I said.

"Oh my God!" he screamed.

He was shocked, like everyone else in the room. They just couldn't believe it. We discussed that we had heard the actual number but in order to respect our discussions with Operation Warp Speed we decided to announce the range of "more than 90 percent." I was surprised when, a few weeks later, Moderna announced their interim results with a specific number, 94.6 percent, which differed from what we had understood the agreement to be with OWS.

Yolanda initiated the disclosure process that had been previously agreed upon. We had to assemble a meeting with the Pfizer executive leadership team, followed immediately thereafter by a meeting with the board. We had to finalize press releases and many other things. I just sat in the room for a few more minutes, completely amazed. My mind raced. What was next? What were the next steps? What would manufacturing and delivery look like? What about the countries that hadn't ordered yet? Everyone would want to have it now. Would we have adequate quantities to supply them? But then I snapped back to the present. Who just sits alone in a room during such a historic celebration? Especially a Greek! I jumped to my feet and went to see Doug.

"Can I call my son to tell him?" I asked Doug.

During all these months, my son, Mois, named after my father, was home from college throughout the pandemic, and had become my constant companion. He was studying electrical and computer engineering at the University of Illinois and was attending online Zoom classes in his room, which was located above my office. He liked coming down to my little office, where he could eavesdrop on all the conversations that he found interesting. He had become my trusted advisor. He would attend almost all the important Lightspeed meetings, sitting in the back of the room, behind the computer camera. He would also listen to many of my calls with the heads of many countries. At the end of each meeting or call, I would ask him what he thought, and we would discuss it for a while. I would also share with him my most private thoughts about this project and particularly my deepest worries and fears. Talking to him helped me structure my thinking and see things more clearly. Doug knew the important role that Mois had played in supporting me all these months, and he cleared my conversation with him.

"Just make sure he doesn't tell any of his friends today, before the news gets out," he said.

I immediately texted Mois a thumbs-up picture.

He replied instantly by asking, "Strong efficacy?"

I replied, "Beyond expectations. Will talk to you tonight." And he sent me a heart emoji!

Yolanda informed me that the meetings with the executive leadership team and the board were set for an hour later, one after the other. Everyone on the board and the executive leadership team was anxious to hear why they had been summoned suddenly on a Sunday evening to an ad hoc meeting. They didn't know that the study was reading out that day, but they knew something big was happening, probably related to the vaccine development.

But before going into these meetings I had to make a special call. In Germany, our partner Ugur Şahin, CEO of BioNTech, was waiting to hear the results. I called him from a small side office near the conference room to tell him the incredible news. In fact, I video-called him so we could see each other. Once connected, we spoke softly, and I could feel the emotions rise again from my heart to my throat and eyes. I sensed the same from him. News of the high rate of effectiveness washed over his face. We were practically in tears. Ugur had always been confident, but this was overwhelming. We spent a few seconds just looking at each other without saying anything but with both of us sensing a lot. Ugur told me, many months later, that our conversation felt like it moved in slow motion. We had taken tremendous risks, and the results were proving us right. After that period of silence, we discussed that we were gearing up for announcements the following day, and we hung up.

Albert Bourla speaking by video call with BioNTech cofounder and CEO Dr. Ugur Şahin as he shared the November 8, 2020, results of the COVID-19 vaccine's efficacy data
Photo courtesy of Pfizer Inc.

Later, we held the executive leadership team video meeting. Thirty minutes after that call, we did the same with the board. We told them all: today we had a conclusive readout of our interim analysis that was positive. We told them that the observed efficacy was overwhelmingly positive, and that the next day we would announce to the world that our vaccine demonstrated more than 90 percent efficacy and that we were going to file in two to three weeks for an Emergency Use Authorization. The joy and relief among all were profound. They all understood that the landscape would be different starting tomorrow.

After the board meeting, I had to make two additional calls. The first was to Dr. Peter Marks, the director of the Center for Biologics Evaluation and Research of the FDA. I gave him a heads-up of what had happened. He was ecstatic.

Dr. Marks told me to file our application for emergency use as soon as we could and that the FDA would review it as fast as feasible.

Then I called Dr. Tony Fauci. The first thing I did was to ask him if he was sitting down. Tony knew that we were expecting our data readout this day, so when I asked this question, he became very anxious. Later, he told me that my question made him think that the news would be either extremely bad or extremely good.

"Yes, I am sitting. Please tell me," he answered.

"Tony, it was positive. There were ninety-four total cases, ninety for the placebo group and four for the vaccine. Ninety-five-point-six percent efficacy."

Tony is usually very measured, but at that moment he became very emotional.

"Albert, this is a game changer," he told me with a trembling voice.

"We will announce tomorrow," I told him.

I was so happy for him. For all these months, he had stood tall and taken so many blows from the White House and the Department of Health and Human Services. At times it felt that he was the only reliable official voice during this crisis, the one person we could all trust. Now he knew that he had the tools in his hands to turn this pandemic around.

6

Past, Present, Future

"He is a wise man who does not grieve for the things which he has not, but rejoices for those which he has."

—Epictetus, AD 50–135

IT WAS ONLY WHEN I got home that evening around 7:00 p.m. that I began to reflect. I rested in my favorite chair, and the silence enveloped me. I began to cry tears of joy. I just sat there for a few moments soaking it all in. My wife, son, and daughter sat next to me. My son took one of my hands into his and my daughter took my other. As I was working from home all these months, they had been present in all the important moments of this project. They'd lived with me through all the good and bad moments, and now they were so proud of what had been accomplished. That night I thought of the nearly 60 million people around the world who had contracted COVID-19, and the more than 1.3 million who had lost their lives. The magnitude of the crisis and our achievement thus far overwhelmed me, and my heart was full of joy.

In that quiet, poignant moment, I also thought of my parents, Mois and Sara Bourla. Over the past year, because racism and hatred were tearing at the fabric of our great nation, I had begun to tell my family's story.

My ancestors had fled Spain in the late fifteenth century, after King Ferdinand and Queen Isabella issued the Alhambra Decree, which mandated that all Spanish Jews either convert to Catholicism or be expelled from the country. They eventually settled in the Ottoman Thessaloníki, which later became part of Greece following its liberation from the Ottoman Empire in 1912.

Before Hitler began his march through Europe, there was a thriving Sephardic Jewish community in Thessaloníki. So much so that it was known as "La Madre de Israel" (The Mother of Israel). Within a week of the occupation, however, the Germans had arrested the Jewish leadership, evicted hundreds of Jewish families, and confiscated their apartments. And it took them less than three years to accomplish their goal of exterminating the community. When the Germans invaded Greece, there were approximately fifty thousand

Jews living in the city. By the end of the war, only two thousand had survived.

Lucky for me, both of my parents were among the two thousand survivors.

My father's family, like so many others, had been forced from their home and taken to a crowded house within one of the Jewish ghettos. It was a house they had to share with several other Jewish families. They could circulate in and out of the ghetto, as long as they were wearing the yellow star.

But one day in March 1943, the ghetto was surrounded by occupation forces, and the exit was blocked. My father, Mois, and his brother Into were outside when this happened. When they approached, they met their father, who also was outside. He told them what was happening and asked them to leave and hide. But he had to go in because his wife and his two other children were home. Later that day, my grandfather, Abraham Bourla; his wife, Rachel; his daughter, Graciela; and his younger son, David, were taken to a camp outside the train station. From there they left for Auschwitz-Birkenau. Mois and Into never saw them again.

The same night, my father and uncle escaped to Athens, where they were able to obtain fake IDs with Christian names. They got the IDs from the head of police, who at the time was helping Jews escape the persecution of the Nazis. They lived there until the end of the war, all the while having to pretend that they were not Jews . . . that they were not Mois and Into, but rather Kostas and Vasilis.

When the German occupation ended, they went back to Thessaloníki and found that all their property and belongings had been stolen or sold. With nothing to their names, they started from scratch, becoming partners in a successful liquor business that they ran together until they both retired.

My mom's story also was one of having to hide in her own land, of narrowly escaping the horrors of Auschwitz and of family bonds that sustained her spirit and, quite literally, saved her life.

Like my father's family, my mom's family was relocated to a house within the ghetto. My mother was the youngest girl of seven children. Her older sister had converted to Christianity to marry a Christian man she had fallen in love with before the war, and she and her husband were living in another city, where no one knew that she had previously been a Jew. At that time, mixed weddings were not accepted by society, and my grandfather wouldn't talk to his eldest daughter because of this.

But when it became clear that the family was going to head to Poland, where the Nazis had promised a new life in a Jewish settlement, my grandfather asked his eldest daughter to come and see him. In this last meeting they ever had, he asked her to take her youngest sister—my mom—with her.

There my mom would be safe because no one knew that she or her sister were of Jewish heritage. The rest of the family went by train straight to Auschwitz-Birkenau.

Toward the end of the war, my mom's brother-in-law was transferred back to Thessaloníki. People knew my mom there, so she had to hide in the house twenty-four hours a day out of fear of being recognized and turned over to the Germans. But she was still a teenager, and every so often, she would venture outside. Unfortunately, during one of those walks, she was spotted and arrested.

She was sent to a local prison. It was not good news. It was well-known that every day around noon, some of the prisoners would be loaded on a truck to be transferred to another location, where the next dawn they would be executed. Knowing this, her brother-in-law, my dearest Christian uncle, Kostas Dimadis, approached Max Merten, a known war criminal who was in charge of the Nazi occupation forces in the city.

He paid Merten a ransom in exchange for his promise that my mom would not be executed. But her sister, my aunt, didn't trust the Germans. So she would go to the prison every day at noon to watch as they loaded the truck that would transfer the prisoners to the execution site. And one day she saw what she had been afraid of: my mom being put on the truck.

She ran home and told her husband, who immediately called Merten. He reminded him of their agreement and tried to shame him for not keeping his word. Merten said he would look into it and then abruptly hung up the phone.

That night was the longest in my aunt and uncle's lives because they knew the next morning, my mom would likely be executed. The next day—on the other side of town—my mom was lined up against a wall with other prisoners. And moments before she would have been executed, a soldier on a BMW motorcycle arrived and handed some papers to the man in charge of the firing squad.

They removed my mom and another woman from the line. As they rode away, my mom could hear the machine-gun fire slaughtering those that were left behind. It's a sound that stayed with her for the rest of her life.

Two or three days later, she was released from prison. And just a few weeks after that, the Germans left Greece.

Fast-forward eight years, and my parents were introduced by their families in a typical-for-the-time matchmaking. They liked each other and agreed to marry. They had two children—me and my sister, Seli.

My father had two dreams for me. He wanted me to become a scientist and was hoping I would marry a nice girl. I am happy to say that he lived long enough to see both dreams come true. Unfortunately, he died before our children were born, but my mom did live long enough to see them, which was the greatest of blessings. I wished they were with me that evening to share in the triumph that the world would learn of the next morning and my role in it, which would not have been possible without their bravery.

The next day, on November 9, 2020, we issued the press release, and Mikael and I made ourselves available to the media for interviews. Before they began, I called Moncef Slaoui, the head of OWS, to give him a heads-up. It was probably six o'clock in the morning, but I knew from previous discussions that he wakes up every day before 5:00 a.m. The previous day I had considered calling him with the news as well, as I had with Peter Marks and Tony Fauci, but I was afraid that the information would leak if I did so. I was not afraid of him leaking the news. I was afraid that he would have an obligation to inform the White House, and I was afraid that the leak would come from there. Moncef was enthusiastic. I could feel excitement and happiness in his voice.

That morning, Sally and Yolanda joined me at my home before daybreak. Yolanda felt it important that someone from my team be with me on the day we shared this monumental news with the world. Sally and Yolanda both live in New York City, so they rode to my home that morning together. Meeting at 5:30 a.m. for the drive to Westchester, they later told me of the excitement and tremendous pride they felt during that car ride. As they watched the sun rise along the horizon, their emotion was overwhelming. COVID-19 was raging, and we were about to share the first news of hope with the entire world.

The good news spread like wildfire. It became the leading story, and in many cases the only story, of news media in every single country of the world. The coverage was amazing. After many months of darkness, the world was facing, for the first time, very positive news that was creating hope. People could see the light at the end of the tunnel. I spent the day giving interviews and talking on the phone with many heads of countries that were calling me to congratulate Pfizer. I took a break later in the afternoon only to attend the regular Project Lightspeed meeting, just as I had done every Monday and Thursday for the past several months. The beginning of this meeting was dedicated to celebrations, but soon we got back down to work. We started focusing on what would be needed to ensure that we would be able to file with the FDA and other regulatory authorities as soon as possible. We also focused on ensuring that those early batches of the vaccine that we'd manufactured at risk would be ready on time so we could send doses out just hours after the expected approval.

In the days following, our news kept monopolizing the press reports. Heads of state kept calling to congratulate me and Pfizer. Among the first was Benjamin Netanyahu, who called me on November 11. I received a lot of calls from US members of Congress, including House Speaker Pelosi, Senate Majority Leader McConnell, Senate Minority Leader Schumer, and House Minority Leader McCarthy. In the meantime, I started receiving news that President Trump was extremely dissatisfied with Pfizer and me personally because the results had come after the November 3 election. He was forming an opinion that this was done on purpose to hurt him and that if we'd wanted, we could have had the results before the elections. The same sources were telling me that Health and Human Services Secretary Azar was thinking the same and was among the people who kept feeding the president's anger. A few days later, I received a call from Vice President Pence. Knowing the atmosphere in the White House, I was afraid that he was calling me to complain about it. But no. In a great demonstration of class, he called to congratulate me and thank me for what we'd done for this country and the entire world, to use his words. He didn't even mention anything else. I was very impressed by the man. President Trump never called me, either to thank Pfizer or to complain.

In a twist of fate, this week also became a week when my personal life entered the spotlight, in an inevitable but unfortunate way. My predetermined plan to sell Pfizer stock at a certain price was triggered when the price of our shares ever so briefly hit a target set much earlier. Most people don't know, but CEOs can't just easily sell their shares of the company they run. Because they are aware of so much material information, they may be faced with allegations that their sale is being driven by inside information that is not available to the public. For this reason, lawyers usually recommend using a plan called 10b5–1. The process of 10b5–1 plans requires that you predetermine the number of shares that you would like to sell for a specified time in the future and the limit price at which you would like to trigger this sale. At Pfizer, we require the extra precaution that the sale can only be triggered at least two months after you have created your plan. This creates a minimum distance of two months between when you set the limit price and when the shares can be sold and helps ensure compliance with the securities rules that prohibit trading when in possession of nonpublic material information. I set this plan originally in February 2020, before we even knew that COVID-19 would become a pandemic or that Pfizer could be working on a vaccine. I used Fidelity as the administrator of the plan, and I predetermined the limit price at $41. In August 2020, before its expiration, I renewed the plan for an additional year with the same predetermined number of shares and strike limit price. After we announced the results of our Phase 3 study on November 9, the stock crossed the limit price at one point during that day, and Fidelity executed the sale automatically. I was not even aware that the sale was executed for an additional day until Fidelity informed me about the transaction. Once this information became public, the same TV shows and newspapers that had been glorifying me two days before started casting doubts, as if my sale were motivated by something I'd known would happen on November 9. I was not used to this kind of attention, and I was devastated. I could feel for the first time in my life the other side of being a public figure, and how fast things can change, even if just two days ago you "saved the world."

In the following days, an army of scientists was tabulating data and preparing all the sections that a submission for approval to regulators (or Emergency Use Authorization in the case of the FDA) requires. I was getting updates twice a day about the progress of the submission. The volume of data required is so large that typically a submission dossier will have thousands of pages. In past decades of vaccine development, pharmaceutical companies sent clinical trial data analysis on paper to the FDA authorities through the mail or overnight courier services. Later, the data was placed on a hard drive that was sent to them, and more recently uploaded to an online portal. This time, Operation Warp Speed activated the US Secret Service. They told us it was too risky to use the portal. We were given an encrypted hard drive, and we were told that the Secret Service would pick it up at one of our facilities outside Philadelphia. We were told that the password at exchange would be "Yellowstone." Our people were working until literally the last moment to download the information onto the hard drive. On the day of submission, Friday, November 20, like a scene from James Bond, two identical solid black SUVs arrived at our building in Collegeville, Pennsylvania. The agents were brought to the room where the hard drives were kept, and they took custody of it. Both parties signed the chain of custody documentation, and the Secret Service asked everyone to leave the room. Then one of them put the drive in one bag he was carrying while another one put an empty hard drive in an identical bag. When they left the room, no one knew which agent was carrying the drive with the information. Then each of the two agents jumped into different cars and drove away. With that mission accomplished, the team in Pennsylvania popped open a bottle of champagne to celebrate. It had been a long week.

The same set of information was submitted to many authorities around the world. The United Kingdom authorized the vaccine on December 2, 2020. This was the first approval of any mRNA COVID-19 vaccine by any country. Emergency Use Authorization for the US was issued by the FDA nine days later, on December 11, 2020. Israel approved it a couple of days later. The European Union approved it on December 21. And the World Health Organization issued its validation on December 31, 2020. In all these cases, the Pfizer/BioNTech vaccine was the first one to be approved by each of these institutions. In the weeks and months that followed, our COVID-19 vaccine was approved and distributed in more than one hundred countries.

7

Manufacturing, the Second Miracle

"Ever to excel."

—Homer, 750–701 BC

DISCOVERING THE VACCINE IN RECORD time felt like a miracle. The second miracle would be our ability to quickly manufacture and distribute it at such scale. Vaccine development and manufacturing are yin and yang: two sides of the same coin. Transforming the formula from an equation in the lab into billions of doses that could be safely transported to millions of locations around the world where trained frontline workers could inject lifesaving vaccine into the arms of eager patients required its own moonshot. In fact, it required a series of smaller, strategic moonshots—a dynamic supply chain, manufacturing precision, and unprecedented logistics over ground, air, and sea. Without world-class expertise at every step of the way, the entire enterprise might have failed.

The secret ingredients in these complex moonshots were the innovation and unstoppable will of our people.

Manufacturing at Pfizer is led by Mike McDermott. Mike had worked in parallel with the R&D team, but after the readout in November 2020, the baton was passed to his team, those who would produce massive doses for the world. Manufacturing at Pfizer is composed of approximately twenty-six thousand people spread across forty-two sites worldwide.

Pfizer Global Supply, which operates sites in China, had become engulfed in the COVID-19 crisis back in January 2020, before the virus spread to become a global pandemic. Mike and his team had to work quickly to ensure they could continue to make medicine there. The next major outbreak was in Italy, where we also operate manufacturing facilities. By the time COVID-19 was first reported in the United States in late January 2020, we had learned a lot about our first priority—the safety of our colleagues in these manufacturing and supply chain settings. The professionals there worked nonstop to ensure that the global supply of medicines continued to be available to patients. In short order, we implemented an unprecedented and comprehensive preparedness plan to control site operations. These actions included reducing the workforce at our sites to only those critical to producing medicines; instituting social distancing and implementing enhanced cleaning procedures for the sites; and expanded use of the Digital Operations Center, which is a Pfizer-developed digital platform that enables our supply operations. The expansion provided Visual Management, a communications tool that allows a snapshot of manufacturing operations, and action-tracking capabilities that enabled colleagues at our sites—either physically or virtually—to stay connected and work collaboratively while social distancing.

Taking every precaution to make our own sites safe and secure, we were able to support the industry collaboration plan I'd outlined in March 2020—making our capacity available to other biopharma companies. Gilead Sciences, for example, used our facilities to meet growing demand for the antiviral remdesivir. Mike's team was doing all of this while also producing medicines essential to ease mounting pressures on hospitals, where patients were going on ventilators and needed our drugs for sedation, muscle relaxation, and pain relief.

But the pressure on manufacturing was just beginning. At first, when the team learned that we would be producing a vaccine against COVID-19, there was tremendous excitement. They'd assumed it would take research and development a year or more to discover the right formulation, and then they would have another year to fill and package the final product. But that was not the timeline I'd envisioned.

I asked our teams to put their best people on the project, parallel track, and not let money be a factor. Normally, we would pursue R&D and manufacturing sequentially, meaning we would conduct the research first, and only if it was proven to be successful would we spend the resources to prepare for manufacturing. With COVID-19, we didn't have that luxury. We did these things in parallel, resulting in a much bigger expenditure of resources. But it didn't mean we would cut corners. Safety and efficacy were paramount. I asked each team to assemble budgets for what they would need. Manufacturing initially projected $500 million just to get started. These projections grew quickly to $850 million. When they presented the number, not only did I tell them that they had that amount, but I also asked them what they would need to move even faster.

For context, total vaccine production at Pfizer at this point, before COVID-19, was two hundred million doses per year, including Prevnar, a pneumococcal flu vaccine to protect infants, kids, and adults. Prior to the pandemic, Prevnar was our largest-volume vaccine, and it had taken ten years to get to that level of production. I knew that we would need to double our overall vaccine production in just nine months. And we would need to do so on an mRNA platform that had not been used at any scale, ever.

What's more, manufacturing did not know which vaccine candidate it would eventually produce. They'd had to build plans for four different formulations with four different potential manufacturing approaches. Normally, they would execute against one construct. Four constructs were narrowed to two before we ultimately chose the b2 formulation. Until then, manufacturing might be working at lightspeed on b1 in the morning before we changed course in the afternoon to b2. John Ludwig, a pharmacist from Chesterfield, Missouri, who leads our medicinal services, reassured his team that if the plan wasn't changing daily, we probably were not doing all we could to end the pandemic.

Constant deadlines and whipsawed priorities provoked frustration, fatigue, and tension. Both site-essential workers and remote workers felt the toll. Site workers were at risk of getting COVID every time they left their homes for work. The fact that they did so on a regular basis to find a source of hope for patients was awe-inspiring.

Pearl River, New York, for example, where we had research labs, was also a community hot spot for the virus. So the people working on the vaccine came into work every day from among a population that was very much in harm's way. More than 350 people a day came into the Pearl River site in the middle of the pandemic. They were diligent with precautions—masks, handwashing, social distancing, protective gear—but they were coming in for long hours and through weekends. It was a tremendous effort under challenging circumstances. During the pandemic, we saw thirty-four hundred Pfizer colleagues infected across the globe. Dozens were hospitalized. As of July 27, 2021, twenty-three colleagues and four contractors had lost their lives. When I would learn of grieving families and families in distress, I would personally call or email.

Still, millions were dying worldwide. We assessed complementary capabilities and capacities across our spectrum of plants and facilities. We went down the menu of sites looking for how to optimize for safety and speed. We have manufacturing and distribution sites across the US, but we landed on a trio of sites upon which we would bet the farm. The first was in St. Louis, Missouri.

The Chesterfield St. Louis site is where the plasmid DNA for the vaccine antigen is produced. The DNA is the template used to manufacture the mRNA sequence included in the vaccine. The template DNA is produced in a cell culture process and subsequently purified through a series of chromatographic and filtration steps. The purified template DNA is then linearized in preparation for the manufacture of the mRNA drug substance at our Andover, Massachusetts, facility.

Using an enzymatic process, the Andover manufacturing facility is the site where the linearized template DNA is incubated with mRNA building blocks in a reaction vessel to make the mRNA drug substance. The mRNA drug substance is then purified to ensure it meets our high standards for quality and subsequently shipped for formulation and further processing to two separate manufacturing sites. The first was in the US, in Kalamazoo, Michigan, the birthplace of William Upjohn, the founder of the Upjohn company, and the location of our largest sterile injectable manufacturing facility in the country. The second was in Europe, in Puurs, Belgium, a little Antwerp town known for white gold asparagus and delicious beer, and the location of our largest sterile injectable manufacturing facility in Europe. I had asked our team to ensure that we would have at least two manufacturing facilities performing every step of the manufacturing process. With this duplication I wanted to ensure that if something happened to one facility, the world would still have the other to continue production. I wanted also to ensure that we had separate lines in case one of them implemented export restrictions, a precaution that later on was proven very wise to have taken.

There the mRNA drug substance and other raw materials are combined through a series of steps, including impingement jet mixing and specialized mixing to construct the all-important lipid nanoparticle, which is followed by sterile filtration. The bulk vaccine was then transferred to an aseptic filling line, where it would be filled into a sterilized vial and capped. There it would undergo inspection before being transferred to the packaging lines for labeling and packing. The packed containers went into storage freezers awaiting final packing into dry ice shipping containers.

The rise of nationalism, however, threatened our goals. Some countries, in an effort to protect local interests, blocked ingredients and substances we needed for the vaccine from leaving their country. We therefore needed a redundant supply chain just in case one country or a block of countries were to lock down. It was a three-dimensional puzzle. Among many other things, we needed glass vials and stoppers to cap the vials. Thankfully, not only are we one of the largest buyers of those, but we also have great relationships with suppliers.

The key ingredient in our COVID-19 vaccine would prove to be lipids. Remember that our vaccine uses lipid nanoparticles to transport mRNA to instruct cells to make the SARS-CoV-2 spike protein. Lipids, which are chemically synthesized, became the most important constraint to solve for. They were a new ingredient, not used at scale in other vaccines.

To make matters even more complex, there are four different lipids needed. Two are proprietary and two are commodity. And so where does one go to buy large volumes of diverse lipids? It turns out, not many places. On the commodity side, niche chemical companies became critical partners. The term "working together" does not sufficiently describe the extreme collaboration we had with the entire biopharma ecosystem. But even that was not enough. We would have to get into the production business ourselves. We started producing our own lipids to increase the availability and ensure production continuity. We started doing this in Groton, Connecticut, and eventually in other places as well. We usually do not produce raw materials ourselves, but this was not business as usual.

It's one thing to make an mRNA lipid nanoparticle in the lab for research purposes. It's quite another to do so at industrial and commercial scale. During the research phase, we designed and built a technical solution consisting of a set of high-pressure pumps, one for aqueous materials and another for organic streams, and a computer mouse–sized contraption we call a "T-mixer." The aqueous and organic formulas are pumped into the T-mixer, and the internal geometry there enables the formation to combine into the lipid nanoparticle. This whole process is controlled by artificial intelligence through a very sophisticated algorithm that was coded specifically for this purpose. The lipid nanoparticle is then purified and ultimately filled and finished into the final product you see in glass vials.

As we set our sights on producing larger and larger volumes of vaccine, we became concerned about the capacity of our already overburdened suppliers. The solution we landed on was to build our own capacity. This meant replicating the pumps and T-mixers dozens and dozens of times into lipid nanoparticle skids. The tech industry has seen warehouse-sized data centers with hundreds and hundreds of racks of network computers. Those racks of computers combine to form cloud computing. It's a similar idea with our lipid nanoparticle skids. Combined, they can produce massive doses of vaccine. We focused on Good Manufacturing Practice (GMP), fabricating the tools and the software we needed. With our newly discovered capacity, Pfizer and BioNTech had shipped hundreds of millions of doses to more than one hundred countries around the world.

Moncef Slaoui, from Operation Warp Speed, toured our Kalamazoo site and said to *The Washington Post* that he came away impressed with Pfizer's commitment. It was apparent Pfizer planned to use its enormous global size, vast stores of cash, and swarms of engineers in a "bulldozer, brute-force" strategy to make billions of mRNA-ferrying nanoparticles, he said.

Mike McDermott and his team were an integral part of our twice-weekly Lightspeed meetings. In time, he championed our reenergized culture to think big and be bold. But there is a difference between thinking big as an aspiration and delivering big on a commitment. As with others on the team, I explored with our manufacturing team the limits of what was possible. At times it was like pulling teeth. There was mutual consternation, but things were moving in an impressive way. In the early days of the crisis, we set two hundred million doses as the annual goal. That number grew quickly to five hundred million. And then I'd ask, why not one billion? When that was accomplished, why not more? Mike told me that what we're doing already is a miracle.

"You are never satisfied," he said. "What you are asking is impossible."

Both statements were true. They had already done more than I could dream when we started this journey. And I was never satisfied. I kept asking for more. But I knew that I had assembled the dream team. These men and women were phenomenal, and I knew we could do it. In fact, I believed that we were the only ones who could do it.

Things were progressing, but there were a lot of ups and downs. And I was relentless in keeping the pressure up. I am very good at remembering numbers. When a projection was presented one week and then lowered without explanation the following week or month, I was livid.

"The timelines on the schedules of slide number seven don't match the dates at the executive summary slide. Your numbers at slide eleven are lower than what you showed me last time. You should have these changes up front and highlighted, not buried on a slide in the middle of the presentation deck. Please don't do this again. You know how well I remember the numbers."

Yolanda finally called members of the Pfizer Global Supply team and advised them to just call me before the meeting if a number was changing. "Wouldn't it be better to share difficult information in a one-on-one setting, rather than on the team call?" she asked. They all agreed that it made sense.

During this process, I saw Mike transforming into one of the best leaders I know. He adopted the "think big" mindset. Once he did this, he felt liberated. He would set bold goals and take ownership of finding solutions for every obstacle that could stand in the way of accomplishing them. I remember vividly when he came to a meeting to present how he would increase production to three billion doses in 2021. One of the other challenges would be that if we needed to increase the production capacity dramatically, we would need many replicas of this new equipment, and we didn't have enough readily available space in our manufacturing sites.

"I will have to build new formulation sites," Mike said, "and as you know, building construction takes years."

And before I could ask him, "What you are going to do about it?" he anticipated my question.

"But we have a solution."

He then showed us a slide with a car assembly line that was operating literally under a tent. It was from a Tesla manufacturing site.

"When Elon Musk told the world that he would build a manufacturing plant for making Tesla cars in a year, no one believed him. But he found a way to place the equipment under very light construction," Mike said.

I was truly impressed by this opening.

"Of course, we are doing aseptic formulation and we cannot host our new formulation suites under a tent. But the concept is what is important here. We can order prefabricated modules that we can install in our Kalamazoo manufacturing site within months, not years. There is a producer in Texas that makes them. We can give him specifications, build them there, and transfer them to Michigan with special large trucks. We will have to do the transfer at night, and we would need the police forces of several states to cooperate with us for this operation, but it can be done. It would take us months instead of years," Mike said.

He continued providing technical details about the fabrication, transport, and assembly of these prefabricated units. I was hearing him talk and was feeling really proud of him. This liberated version of Mike was the brilliant leader that Frank D'Amelio and I had seen when we appointed him as the head of our manufacturing group.

In early 2021, we committed publicly to making 2.5 billion doses by the end of the year. The manufacturing team surprised themselves and took a lot of pride in their success. If I've learned anything during my twenty-eight years with Pfizer, it is that people will tend to underestimate what they can achieve, and in an organization like ours, the people below them will underestimate a little more. You have to extend yourself. You can do more than you think you can. People can push themselves more than they think possible.

Manufacturing the vaccine had left us with a product as fragile as a snowflake. We now had to deliver that snowflake with great care around the world, which we would accomplish using boats, trucks, and planes.

Getting the product right in the early phases had kept me awake at night, and the distribution kept me pacing the floors at all hours. Our mRNA vaccine had to be stored and shipped at an ultracold temperature: minus-seventy degrees Celsius (negative ninety-four degrees Fahrenheit). On the moon, at night, maybe that's no problem. On Earth it's a challenge. Shipping a vaccine at this temperature at scale had never been done before. There was simply no infrastructure to support it. Imagine having a rocket—the vaccine—with no place to land it. Our global supply chain team came up with a brilliant solution.

Led by a man who became known as the Ice Man, James Jean, our engineers designed a temperature-controlled thermal shipper that could transport and store the vaccine anywhere in the world. We had been piloting the idea before COVID-19, but when the pandemic struck we skipped the test phase and went straight to full implementation. The shipper, about the size of a carry-on suitcase, weighed about seventy-five pounds. It carried a minimum of one tray of vaccine vials and as many as five trays. Each tray had 195 vials, 6 doses per vial. So a single shipper could carry as many as 5,850 doses.

We captured three readings—location, temperature, and light. GPS kept us informed in real time of the shipper's location, and a thermometer kept track of temperature fluctuations. The light sensor notified us if the shipper was opened or cracked, which was a breach of security. To monitor all of this data, imagine a computer dashboard with a map of the world.

At any moment, we had three thousand or more shippers in the air, on water, on the road, or in a healthcare facility. Our global supply chain team could click on any individual shipper and see its current temperature and location as well as the milestones it had achieved and any exceptions it had logged. For example, our algorithm captured if it reached certain airports on time. The truck carrying the shipper might have been diverted from the planned route due to weather or an accident. If someone opened the shipper, we would know that. We could also see data for the shipper sitting right next to it, so we'd notice any differences between shippers on the same vehicle. We built the shipper to be independent from the shipping company carrying it. FedEx, UPS, and United Airlines received the same alerts and notifications we did, but we were never dependent on a single carrier. Separately, we had a global security operation team that notified us of potential or real threats to the shipment from weather and political or social unrest. In the early days, while we were testing the system, we had an open line at all times for our logistics experts to communicate and collaborate. If we discovered an issue en route, we could stop the shipment and replace it with a new delivery if necessary.

The system was so intelligently designed that our shipment accuracy was 99.9 percent. This was a system that had not existed in what we might now call the pre–Ice Age. Now the shippers can even serve as a supercold storage unit in places that never had that before. In the past, vaccine shippers gathered data, but we had to download that data when the journey was complete and then look backward. Now we look at critical data in real time so we can react, correct, and deliver.

This breakthrough in logistics will become increasingly useful in the future as we explore new mRNA technologies.

Vaccines have always faced a last-mile problem. We can ship the vaccine thousands and thousands of miles by planes, trains, boats, and automobiles, but that last mile into the center of a city or a remote community—all the while preserving a subfreezing cold chain—requires solving a snarl of complications. Inventing a therapeutic drug or a vaccine is the first mile, but there remain many miles to go. Medicines have always faced this challenge, which is why some companies and philanthropies have invested in mobile refrigeration. Cool.

What happened next, in the days leading up to the first shipments of vaccine, will remain with me always. A quarantine-weary world was suddenly fascinated by logistics—shipping, storage, distribution, and use. Like armchair generals during a world war, people studied intricate flowcharts indicating the "movement of troops" published in national and local newspapers. In the US, maps with arrows, diagrams, and datelines illustrated the vaccine's path from our manufacturing and distribution facilities in the Midwest to their state and their local clinic or hospital. Similarly, Europeans followed shipments from our facility in Puurs, Belgium. Some charts began at Pleasant Prairie, Wisconsin. Others started at our plant in Portage, Michigan, just outside Kalamazoo, where the first 2.9 million shots were shipped from our freezer farms to more than six hundred sites. At that plant in Michigan, we had built a freezer farm larger than a football field, capable of holding 100 million doses at any time.

In addition to the freezer farm in Puurs and two others, worldwide storage capacity is 400 million doses. Pfizer employees reported that they could hardly walk to the end of their cul-de-sacs because neighbors would come out of their houses to say thanks and to ask questions.

The first patients to receive our Pfizer vaccine against COVID-19 were at University Hospital in the Midlands of England, north of London. Following Margaret Keenan, the second in line was none other than William Shakespeare, then an eighty-one-year-old patient in the frailty ward. It is only appropriate that the namesake of one of the greatest writers of tragedies would be among the first to receive hope. Hours earlier, our workers had waited anxiously to load boxes onto awaiting trucks as the UK government authorized not just the vaccine but each batch. Truck drivers reached their hourly limits, and replacements were brought in to stand by.

It had been 269 days since we started the vaccine's development.

On Saturday morning, December 12, 2020, *The New York Times* summed it up well: "The authorization set off a complicated coordination effort from Pfizer, private shipping companies, state and local health officials, the military, hospitals and pharmacy chains to get the first week's batch of about three million doses to healthcare workers and nursing home residents as quickly as possible, all while keeping the vaccine at ultracold temperatures." Inside on the *Times*' business page: "An Air Rescue's Goal: Billions of Doses."

On Sunday morning, December 13, trucks backed into the shipping docks at our Michigan plant and hauled away a lifesaving vaccine. The pandemic meant there would be no situation room where we could all gather to witness this historic moment. Our team was scattered in living rooms and breakfast nooks everywhere. Like everyone, I monitored events on television and through our own channels.

Receiving the news of a safe and effective vaccine was very emotional, the best day of my professional life. Watching the vaccine begin its journey to your healthcare provider and your arm at a clinic or hospital somewhere in the world gave me a feeling of completion. Seeing those trucks drive away created joy. Hurdlers and marathon runners experience the same elation when crossing the finish line.

That week in Seattle, where America had experienced its first suspected COVID-19 death nine months earlier, exhausted healthcare workers were at the breaking point. Healthcare workers had worked nonstop; some had been driven from their chosen profession. Hospital beds were either at or near capacity. The lead story in Seattle's local paper quoted Dr. Fauci as saying: "The cavalry is coming." Not long after, a heartbreaking photo of a doctor hunched over, crying after receiving the vaccination, was spread across the front page.

Getting a vaccine from the laboratory to the manufacturing plant to the arms of patients around the world is a relay race. We hand off the baton to local hospitals, clinics, and pharmacies, and they take ownership of administering the vaccine, arm to arm. Take CVS, for example. By April 2021, the local US retail pharmacy had vaccinated ten million patients at 2,100 sites. Setting up test sites early in the pandemic prepared CVS well for the vaccination phase. They hired nurses and technicians to ensure adequate capacity. And thanks to CVS's digital and human outreach, 94 percent of patients returned for their second dose of vaccine on the day of their appointment. If patients had trouble getting to the store, CVS had an arrangement with the ride-share company Lyft to get them there. In a pandemic, we are all only as protected as our closest neighbors.

Another challenge we recognized early on was to reduce waste and increase the number of doses per vial.

At the time we submitted our first regulatory applications, we knew that we were filling our vials with enough vaccine volume so that after dilution each vial would have approximately 2.25 ml. For each dose we needed 0.3 ml, so inside every vial we had volume for at least six doses. The challenge was that you cannot use all the volume all the time. After injection, some volume will remain unused, typically in the space between the needle and the tip of the syringe, as "dead volume." There are many different types and brands of syringes that are commercially available, and each one has a different dead volume. Typically, the issue can be resolved by overfilling the vials to cover all options. But in the case of a COVID-19 vaccine in the middle of a pandemic, every single drop saved could mean additional lives saved. To be accurate, each dose is 0.3 ml, so with any excess ml per vial we had potential to save an additional life. We didn't have the luxury of wasting any doses.

We discussed this issue extensively during our Lightspeed meetings. We were working hard to build incremental capacity, so the fact that we were having such a high waste was clearly not feeling right. The first person who brought this to my attention was Ugur Şahin from BioNTech. He had noticed that we were significantly overfilling our vials and told me that according to his calculations we were wasting 40 percent.

"Albert, we must find a solution to this," Ugur told me.

I brought it up to the team, and we started exploring solutions. We agreed to a plan that would allow us to test many different combinations of syringes and needles and find how many doses could be extracted with each. But at the time of the submission, we hadn't done the work to determine the maximum number of doses that we could safely extract from each vial, and for this reason, we filed our application indicating only five doses per vial.

Following our global submissions, our vaccine was approved in many countries of the world as five-dose vials, but in the meantime we continued working to gather the evidence for higher doses per vial. It felt like we sampled every syringe and needle combination in the world. In December 2020 and January 2021, our country managers called local manufacturers and we created a list. Then we tested them all. We needed syringe and needle matches that would extract and dispense six doses consistently. We discovered a very large number of syringe and needle combinations that could achieve that, and by using the right combination of low-dead-volume syringes and needles, we discovered it was possible to reduce the waste and get safely at least 20 percent more vaccine out of each vial. But there was an even bigger obstacle to overcome. There simply were not enough of the necessary syringe and needle combinations to serve the world. We'd have to invent that supply chain as well. It became a new, parallel work stream. We began by activating another team of people to make it happen. Initially, we needed a billion syringes and needles. Fortunately, our suppliers were eager to help.

We shared our production plans with a medical tech manufacturer, and they activated their facilities and ramped up production. In some cases, manufacturers wanted financial guarantees that this large number of syringes and needles would in fact be purchased by providers on the ground. We told these manufacturers to go for maximum production without worries. Pfizer would buy every syringe and needle they couldn't sell.

It worked. They all started maximum production immediately and gradually flooded the market with this type of needles and syringes. We then had to get the regulatory approval for the change. Once the data supporting six doses was available, we immediately submitted it to regulatory authorities around the world. The six-dose label was first approved by Israel on January 3, 2021, and then by the other regulatory agencies in short order, including by some of the most renowned regulatory agencies in the world, such as the US FDA, the

European Medicines Agency, the UK Medicines and Healthcare products Regulatory Agency, Switzerland, and the World Health Organization. The next step would be to retrain those administering the vaccine to make the pivot from five to six doses per vial so we could get it right for everyone.

With more and more experience, some countries are now seeing up to seven doses per vial, which has raised the overall average to near 6.6 doses per vial.

Inside Pfizer, I knew that colleagues would feel pride about what we had accomplished together. But I didn't fully appreciate how great that pride would be. Mike McDermott's five daughters drew him a bright, cheerful poster that said, "Thank You," with a hashtag for his division—#PGSProud. One group opened an online company store offering Pfizer T-shirts and other items with our logo and the tagline that came to represent our commitment to science: "Science Will Win." So many employees went at once to buy items for their family that they crashed the website. Our Chief Human Resources Officer, Payal Sahni Becher, told me she wouldn't be surprised if some employees got a Pfizer tattoo. I would not be among them, although I have to admit I thought about it for a moment.

Looking back, I had a personal scare in those last days before we knew the effectiveness of the vaccine and before we ramped up manufacturing. In early December, I was in contact with two people who later tested positive for COVID-19. They visited the emergency room but were not admitted. We learned of their COVID-19 infection late in the afternoon on December 11, the same night we were expecting to receive news of the vaccine's Emergency Use Authorization from the FDA. The irony of it all was quite dramatic. Here I was, on a momentous evening, and I learned that I had been directly exposed. My team, suddenly in crisis mode, moved quickly to find a way to have me tested. In three days, I was scheduled to do an in-person interview with *60 Minutes* at our Pearl River site, and it was important that I not inadvertently expose anyone while there.

I discussed possible options with Yolanda and Payal. I could visit my personal physician, but it was 4:30 on a Friday afternoon, I knew there was no way he was available. We also discussed the possibility of me going to a local testing center, but my team did not like that option at all. At that time, the lines at local drop-in centers were blocks long. The notion of me, the CEO of Pfizer, caught on video waiting to be tested on the night we were expecting to hear from the FDA about our vaccine did not seem like a very good idea to anyone. Fortunately, we came up with an alternative plan to test me, and the results came back quickly and were negative. I had not contracted COVID-19.

By mid-February 2021, I had a chance to celebrate in a historic way the accomplishments of research, manufacturing, and first-line workers. It was midwinter 2021 when I received a top-secret phone call from the White House asking if President Biden, less than a month after his inauguration, could join me for a tour of our Kalamazoo manufacturing site, which is actually located in the small town of Portage, Michigan. The Secret Service asked me to keep it quiet, but I asked for and received approval to call Mike McDermott.

"Mike, we can't say anything about this to anyone, but President Biden would like to visit our plant in Kalamazoo next week. How should we do this?"

"We will tell everyone that you are coming to inspect, and they should start preparing. Setting aside the Secret Service, we wouldn't do that much more to prepare for the president."

It felt strange hearing this. I remember thinking, *My God. Am I really causing that much disruption when I visit a Pfizer site?* Now I understood why Mike had refused to have me visit our manufacturing site in Andover at the beginning of the pandemic by telling me I was not "essential."

The tour was set for Thursday, February 18, but a snowstorm in the upper Midwest threatened to suspend the presidential visit. I was prepared to be disappointed, but the White House called again to ask if the president could come the next day, Friday, instead. Our immediate answer: Of course.

When the motorcade arrived, it was a patriotic moment for me to see the president of the United States, followed closely by military personnel carrying the "football," with the nuclear codes, at our company's facility. He congratulated everyone at Pfizer, and together we made our way from machine to machine, where workers greeted us and explained their role and how the technology worked. It was amazing to see the president's personality simply and easily become that of one of them—a blue-collar worker. He was obsessed with their stories and their jobs. He was clearly in his element. We were running late to address workers and the national media when he pulled me aside, placed both hands on my shoulders, and looked me in the eyes. "I hear your parents survived the Holocaust," he said. I sensed his staff growing anxious for us to proceed onto the stage. He ignored them. He began to tell me about other families he had known who still bore those scars. The world grew quiet for just a few moments, and I could feel tears in my eyes and the weight of his embrace on my shoulders.

Only in America.

8

Equity: Easier Said Than Done

"The way to gain a good reputation is to endeavor to be what you desire to appear."

—Socrates, 470–399 BC

Wedding photo of Mois and Sara Bourla. At the far left is Josef Saias, and at the far right is Miko Saias, Albert Bourla's maternal uncles who survived Auschwitz. Behind Sara Bourla to the right is her sister, Freda Dimadis, whose husband, Kostas Dimadis, bribed the German Commander Max Merten to spare Sara's life when she was arrested by the Nazis. *Photo courtesy of the author*

Albert Bourla blowing out the candles for his third birthday. His mother, Sara, is on the far left and his father, Mois, is in the back middle. Albert's sister, Seli, stands next to him. The other children are Albert's cousins. *Photo courtesy of the author*

Pfizer's Purpose Circle, created by Albert Bourla to cultivate a sense of equality in team discussions. The photos of Pfizer colleagues' family and friends—real patients—are displayed on the walls to remind them of Pfizer's purpose: Breakthroughs that change patients' lives.
Wendy Barrows

Left to right: Pfizer's General Counsel Doug Lankler, Chief Scientific Officer Mikael Dolsten, CEO Albert Bourla, Chief Corporate Affairs Officer Sally Susman, and Chief of Staff to the CEO Yolanda Lyle upon learning of the extraordinary efficacy of Pfizer's COVID-19 vaccine, November 8, 2020.
Photo courtesy of Pfizer

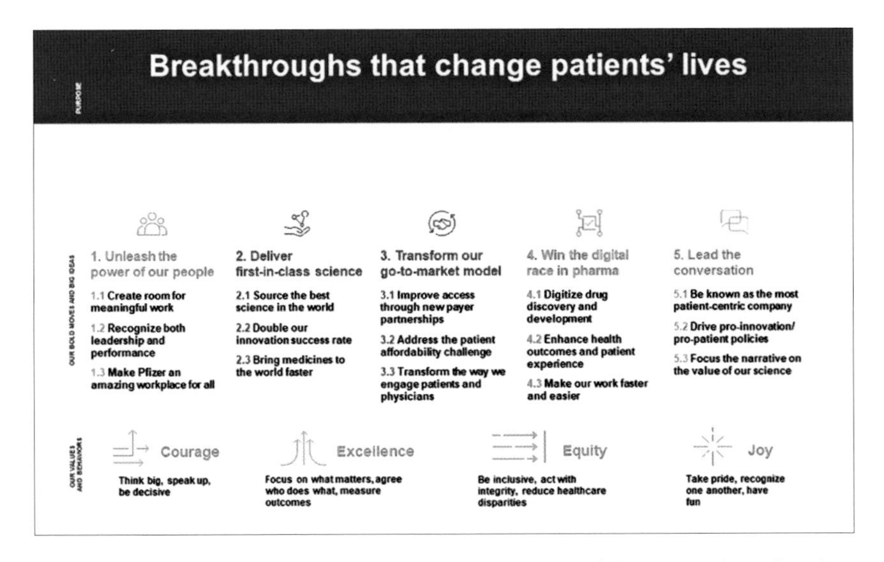

Pfizer's purpose blueprint and values, introduced by Albert Bourla in his first year as CEO of Pfizer. *Image courtesy of Pfizer*

First Pfizer COVID-19 vaccine shipments departing Brussels, November 30, 2020. *Photo courtesy of United Airlines*

Pfizer colleague preparing the first shipment of COVID-19 vaccines out of the Kalamazoo, Michigan, manufacturing plant, December 13, 2020.
Photo courtesy of Pfizer

U.S. President Joe Biden and Albert Bourla in conversation following the Kalamazoo manufacturing plant site tour, February 19, 2021.
Photo courtesy of Pfizer

Albert Bourla meeting with Albanian Prime Minister Edi Rama at Pfizer headquarters, December 9, 2020. *Photo courtesy of Pfizer. The Andy Warhol Foundation for the Visual Arts, Inc. / Licensed by Artists Rights Society (ARS), New York, © 2021*

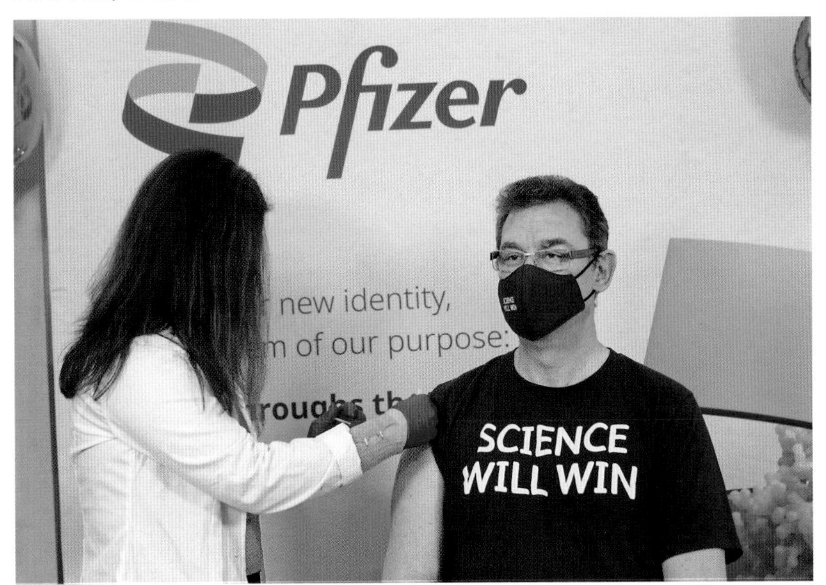

Albert Bourla receiving his first dose of the Pfizer COVID-19 vaccine at the company's Pearl River site, February 16, 2021. *Photo courtesy of Pfizer*

Left to right: Belgian Prime Minister Alexander De Croo, BioNTech cofounder Dr. Özlem Türeci, Puurs Site Lead Luc Van Steenwinkel, European Commission President Ursula von der Leyen, Pfizer CEO Albert Bourla, and Pfizer Puurs Experimental Pilot Plant colleague Marjoh Nauta on a tour of Pfizer's Puurs, Belgium, COVID-19 vaccine manufacturing plant, April 23, 2021. *Photo courtesy of Pfizer*

Albert Bourla with Executive Leadership Team members and Pfizer Occupational Wellness colleagues immediately after receiving their COVID-19 vaccines at the company's Pearl River site, February 16, 2021. *Photo courtesy of Pfizer*

UK Prime Minister Boris Johnson shared this photo on his LinkedIn account following the conversation he had with Pfizer leadership about the COVID-19 vaccine, January 14, 2021. *Pippa Fowles / No 10 Downing Street, 2021, licensed under CC BY 2.0*

Political cartoon that appeared in Hebrew in the Israeli newspaper *Haaretz* on May 30, 2021. *Cartoon by Eran Wolkowski,* Haaretz

Albert Bourla and President Joe Biden at the G7 Summit in Cornwall, England, as they announce the U.S. government's donation of 500 million doses of Pfizer's COVID-19 vaccine to low- and middle-income countries, June 10, 2021. *Brendan Smialowski / AFP via Getty Images*

WHEN WE'D STARTED THE COVID-19 project, I had made clear that return on investment should not be of any consideration. It was already May, and still we had never spoken of a price for the vaccine or calculated revenue streams. But several governments had started procurement discussions, and they all wanted a price. It was time to make a decision.

The way we price our medicines is by calculating the value they bring to patients, to the healthcare system, and to society. Unlike what some may believe, good medicines reduce, rather than increase, the cost of a healthcare system. We try to calculate this economic value. For example, if one hundred people take a heart medicine and as a result we have five fewer heart attacks, we calculate the cost that these five heart attacks would generate to the healthcare system (ambulance rides, hospital stays, tests, doctors, caregivers, work days lost, etc.) and compare it to the cost of the medicine for one hundred people. Of course, there are many more nuances to this economic value. How does one put a price on the avoidance of human pain? This, in my opinion, is truly invaluable.

I asked our pricing team to calculate the usual economics of the global COVID-19 crisis, and they came back with staggering numbers. For an assumed 65 percent efficacy, the reduction of hospitalization costs alone would be hundreds of billions of dollars. We could price the vaccine at $600 per dose, and still the healthcare system would pay less than it saves—not counting the value of human lives saved. I realized that this could become a gigantic financial opportunity for us but also that in the middle of a pandemic we could not use the standard value calculation for setting the price. I asked for a different approach. I told the team to bring me the current cost of other cutting-edge vaccines like for measles, shingles, pneumonia, etc. In the US they were priced between $150 and $200 per dose. It sounded fair to me to match the low end of the already existing vaccine prices. No one could say that we were using the pandemic as an opportunity to set prices at unusually high levels. I told my team to begin procurement discussions at this starting point and offer discounts for volume commitments. The UK and US governments as well as the European Union had already approached us for procurement negotiations, and we had started preliminary discussions with them.

But soon thereafter, and while these discussions were ongoing, a level of discomfort started gnawing at me. I was thinking that we might be missing an opportunity to gain something more valuable than a fair financial return. We had a chance to gain back our industry's reputation, which had been under fire for the last two decades. In the US, pharmaceuticals ranked near the bottom of all sectors, right next to the government, in terms of reputation. I again asked the pricing team to give me the current prices of the cheapest commodity vaccines. In the US, flu vaccines cost up to $70, but they also offered a low protection rate of around 50 percent. Their low end is around $20 to $30.

"We are changing course," I told them. "For the high-income countries, the starting point should be the low end of flu pricing. We can still offer discounts for high-volume commitments."

"But this is the cost of a simple meal, not a cutting-edge vaccine," someone replied from the pricing team.

The cost of a simple meal. I liked this expression very much. Later, I would use it often when asked by journalists about pricing. I turned to the colleague who had told me this and replied:

"I like that. You could make a great career in marketing."

We went back to the US and European negotiators and reduced our price significantly. They were very positively surprised. But the vaccine had to be accessible to everyone, everywhere, in an equitable way.

Equity doesn't mean that we give everyone the same. Equity means that we give more to those who need more. Therefore, we couldn't have a single price for all. Instead, we decided to implement a three-tiered pricing approach. We used the World Bank's classification of economies for our own analytical purposes. The World Bank uses gross national income (GNI) per capita data in US dollars, converted from local currency using the World Bank Atlas method, which is applied to smooth exchange rate fluctuations, to classify countries into four categories: high-income countries, upper-middle-income countries, lower-middle-income countries, and low-income countries. We had just agreed that the price for the wealthier nations—the high-income countries—would be the "cost of a simple meal." The upper-middle-income countries would be offered vaccines at approximately half of this price. And the lower-middle-income and low-income countries would receive the vaccine at a not-for-profit price. The only stipulation was that the countries make the vaccine free to their citizens.

"In pandemics you are only as protected as your neighbor. And it is extremely important that we not let price be an obstacle to anyone. Not only because it is the right thing to do, but also because it will be a threat—not only to these countries, but to all," I said.

The team was very happy with the clarity and equity we were bringing to our approach. The next day they prepared pricing guidelines that were sent to all our local operations in every country with the instruction to approach their local governments and ask for advanced purchase orders under these conditions.

Angela Hwang, Pfizer's President for the Biopharmaceuticals Group, led that effort with her talented team in regions around the world. A dynamic leader with a long career at Pfizer, Angela's childhood experience was that of an Asian girl growing up in apartheid-era South Africa. This defined her worldview and underscored the importance of equity.

Angela and her team set out to sign up every country to order the vaccine for its citizens in one of the three pricing tiers. But they also helped the country's frontline workers to be prepared. The education effort was enormous. In the United States alone, they worked with the states and Operation Warp Speed to train forty thousand people to handle and administer the vaccine. Once a country had the vaccine, it had to know how to store and dilute it, and monitor patients for side effects afterward.

Most of the high-income countries were among the first to place orders to reserve doses of our vaccine through 2021. Europe, the US, Japan, and the UK were among the many that included us in their bets. Unfortunately, many other countries, particularly middle- and low-income, decided to go exclusively with other vaccines, either because mRNA technology was untested at that time or because other companies had promised local manufacturing options. Our local teams tried hard but were unsuccessful in changing leaders' minds. As I looked at my Excel spreadsheet, the disproportionate allocation of doses to higher-income countries became a concerning situation. I made a personal effort to convince some of the middle- and low-income countries that had not placed orders. I sent letters to their leaders, and our local teams followed up, but we were again mostly unsuccessful. In October 2020, even after it had become clear that Pfizer would be the front-runner in quickly bringing a vaccine to patients, a few additional countries had added their names to the list of nations ordering our vaccine, but still not enough low- and middle-income countries to balance inequities.

The situation started to change as soon as the vaccine was pronounced effective. First, the US government, which had ordered one hundred million doses, approached us and asked to order an additional one hundred million. A few months back, we had asked OWS to order these additional one hundred million doses because our supply was about to sell out. They refused. I personally called the OWS leadership and told them that Europe had ordered two hundred million doses, and it would be prudent for them to do the same. I remember saying it would be embarrassing for the US to not be able to get doses from an American company, just because they didn't order enough, and emphasized that we intended to stick to the principle of allocating our production strictly to confirmed orders, respecting our commitments. This was to say that if they changed their minds later, we were not prepared to take doses away from other countries that already had placed orders. OWS had refused again at that time. Now they wanted the additional one hundred million doses, and they wanted them immediately. The problem was that all our projected supply for the first six months had already been allocated to other countries, and it would be difficult to deliver additional doses before June. But we could deliver later in the year. However, the discussions were stalled because of other reasons. That's when President Trump's son-in-law and advisor, Jared Kushner, called me to resolve the issue.

We'd thought that providing the US with an additional one hundred million doses would be straightforward. Our existing contract gave them the opportunity to purchase additional doses, but bureaucracy was getting in the way. Because the vaccine was now approved, they couldn't or didn't want to use the original contract under which they had purchased the vaccine when it was a candidate for approval. I didn't want to reopen negotiations. I felt it would be a Pandora's box, because we had spent weeks negotiating the first contract. In the meantime, criticism of the Trump administration for not having purchased enough doses was mounting. Jared called to understand what the issues were. I explained the situation with the request to start negotiations again with new language in the contracts.

Jared indicated that he thought the bureaucracy was ridiculous and promised to call the right people.

Jared's intervention helped us reach a resolution, and a few days and several phone calls later, our lawyers had come to a mutually agreed-upon solution. But then the issue of the delivery schedule came up again. Jared was asking for a very aggressive delivery plan to the US for the additional one hundred million doses. He wanted it all in the second quarter of 2021. To do that, we would have had to take supplies from Canada, Japan, and Latin American countries, all of which had placed their orders earlier than the US and were expecting the vaccine in the second quarter. I refused to do that, and the debate between the two of us became heated. I reminded Jared that I had made very clear to Moncef Slaoui that we would not take doses from other countries to give to the US and that I had almost begged OWS to increase their ordered quantity in the first contract, but that they had repeatedly refused to do so. But Jared didn't budge. In his mind, America was coming first no matter what. In my mind, fairness had to come first. He insisted that the US should take its additional one hundred million doses before we sent doses to anyone else from our Kalamazoo plant. He reminded me that he represented the government, and they could "take measures" to enforce their will.

"Be my guest, Jared," I replied. "I prefer to have Japan's prime minister complaining to you about the cancellation of the Olympics rather than to me."

Thankfully, our manufacturing team continued to work miracles, and I received an improved manufacturing schedule that would allow us to provide the additional doses to the US from April to July without cutting the supply to the other countries. It was a good compromise, and eventually the contract was signed. Jared called me two days later from Mar-a-Lago to thank me for the collaboration, and we closed the loop on a happy note.

A month later, the same scenario was repeated with the EU. The European Commission had started getting a lot of pressure from member states to implement an export ban for vaccines manufactured on European soil. AstraZeneca failed to supply, and this created a political problem for European leaders who had placed significant bets on them.

The Europeans started implementing an export control system that required us to submit a lot of information before we could export any doses outside the EU. Although the system was not an export ban, it became a significant administrative burden for our Belgian manufacturing site. The complaints kept coming my way. "Do something about it, boss. The people who are working to send the vaccine to so many countries are the same people pushing this additional paperwork. They are exhausted, and we can't train others on the fly."

I called Ursula von der Leyen, president of the European Commission, to ask her to relax these requirements. My pitch was that, as long as we met the delivery schedules, we should be unburdened from this obligation. Ursula insisted on keeping the controls in place, and she pointed out that half of our output from European manufacturing sites is "allowed" to be exported outside the EU while none of the US-produced output was exported outside the US. I knew she was right and acknowledged it. Under the Defense Procurement Act (DPA), which was incorporated into our agreement with the US government, we risked not only civil but also criminal prosecution if we exported the vaccine outside the United States. So, while the US government did not ban exports of the vaccine per se, the repercussions for us doing so could have been dire under the strict guidelines of the DPA. I asked if President von der Leyen was familiar with these details. I asked her to at least reduce the amount of the requested information that we had to submit every time we were sending a shipment to a country. She promised to have a look into it. I knew she would because she was someone who always kept her word.

I got to know President von der Leyen well through our discussions during the COVID crisis. We first spoke on the phone on January 5, 2021, regarding dose allocation. She is a remarkable woman. Born in Brussels, she grew up in a family devoted to public service and went on to serve in German chancellor Angela Merkel's cabinet, including as defense minister. Over the coming months, I would build a close relationship with her through texts and phone calls about the vaccine, virus variants, and manufacturing. She was knowledgeable about all these topics and never missed an opportunity to press for an accelerated supply timetable. And she had a remarkable ability to get what she wanted.

In the earliest days of COVID-19, the World Health Organization (WHO) and partners launched the Access to COVID-19 Tools Accelerator, known as ACT-Accelerator. It was not a partnership, but rather a platform for collaboration across different sectors. There are four pillars of activity, with one pillar focused on the procurement and equitable distribution of vaccines through a facility called COVAX. The goal of COVAX is to accelerate the development, production, and equitable access to vaccines. COVAX is co-led by several global health organizations, including the WHO; the Coalition for Epidemic Preparedness Innovations (CEPI); Gavi, the Vaccine Alliance; and a delivery partner, the United Nations Children's Fund (UNICEF). Its goal is to provide equal access to COVID-19 vaccines for all countries, regardless of income levels. The Gavi COVAX Advance Market Commitment, a financing instrument designed to support the ninety-two lower-middle and low-income economies, was also an important tool to ensure that developing countries have the same access to vaccines as the rest of the world.

Historically, the relationship between the public sector and the innovative pharmaceutical companies had been strained. In particular, the WHO had a contentious relationship with the pharmaceutical industry and strict rules governing how the WHO and the industry could work together. But in 2017, as a new WHO director general was elected for a five-year term—Dr. Tedros Adhanom Ghebreyesus, the first from Africa—a new vision and understanding of public-private partnerships was brought to the WHO. Born in Eritrea, Dr. Tedros served as Ethiopia's minister of health, leading a comprehensive reform of the country's health system, built on the foundation of universal health coverage and provision of services to all people, even in the most remote areas.

As head of the WHO, he proclaimed, "Our vision is not health for some. It's not health for most. It's health for all: rich and poor, able and disabled, old and young, urban and rural, citizen and refugee. Everyone, everywhere."

I was preparing to become CEO when I heard his proclamation, and I remember thinking, *This is exactly what my vision for Pfizer is*, and I couldn't have said it better.

My team warned me that working with the WHO had been tough in the past, but I saw Dr. Tedros's mission and our values as very much aligned. In August 2020, in the middle of the pandemic and with a potential business trip to Europe approaching, I asked my team to request a meeting with him. I was told that Tedros would not likely accept a meeting with a biopharmaceutical company's CEO. He preferred to work through industry associations. I was seriously disappointed by this news. What a wasted opportunity to do something together for the world. How could we let prejudice or preconceived notions stand in the way of us talking? And suddenly, on the morning of Sunday, December 13, I received a WhatsApp text that said, "Dear Albert. I hope all is well. This is Tedros from the WHO. I would appreciate it if I can have a few minutes of your time to discuss COVID. Please let me know the openings you have, and I will arrange. My best, Tedros."

Knowing what my team had told me, I was pleasantly surprised to see this text. I replied immediately, and we spoke the next day. The first thing I told him was how happy I was that he reached out, that the WHO was critical and that I was very disappointed when President Trump withdrew the US from the WHO. I told him I was looking forward to working together to find solutions and ensure equitable access to vaccines and treatments.

The discussion went very well, and I felt that Tedros was deeply committed to the most vulnerable. Because I felt such a connection to him, I couldn't help myself and in the end told him my frustration about having wanted to have this discussion four months earlier, when I visited Geneva, but having been told that he wouldn't take a meeting with a pharma CEO. "What are you talking about?" he replied. "My office is open for everyone who wants to see me." I believed him, and I thought to myself that it was our own preconceived notions that had prevented that meeting in August from happening.

This first exchange would set the stage for a productive and very warm relationship. Not only did we remain in touch over WhatsApp, but our staff did as well. His team supported us to get an agreement with COVAX. So, when Pfizer became the first company to launch a COVID-19 vaccine, our mutual desire for everyone on the planet to get the vaccine led to tremendous collaboration, and the first fruits of this partnership eventually came together. But it was not easy.

I told COVAX leadership that Pfizer was offering the vaccine at cost for lower-income countries, yet they continued to focus on the price for wealthier and middle-income countries; from what I could tell, their allocation methodology did not prioritize lower-resourced countries. It appeared that they wanted to create a global vaccine hub, situating the WHO at the center of that hub. But this was a design flaw. Rich nations could come directly to the vaccine manufacturers and negotiate. It was a matter of national security for those countries. But the poor countries needed COVAX. The aspiration of COVAX did not meet with the reality of how access would unfold for the world. I believe that by focusing on all countries, COVAX missed the point that "equity" doesn't mean doing the same for all but rather means doing more for those who need more. And who really needed their support were the lowest-income countries.

In January, Pfizer and BioNTech signed an advanced purchase agreement with COVAX for up to forty million doses of our COVID-19 vaccine in 2021. Frankly, I was surprised when COVAX sought only forty million doses, clearly not enough to meet their ambitious goals. I asked, "Are you sure? How will you meet the projections you've set?" I was told COVAX had decided to work with AstraZeneca as the backbone of their strategy. They used Gates Foundation funding to build a technology transfer scheme with the Serum Institute of India, a large vaccine manufacturer. But that plan didn't work well. South Africa dropped AstraZeneca because its vaccine did not block the South African mutation, and AstraZeneca couldn't manufacture the volumes they had promised. To compound problems, in May 2021 many nations began to seal their borders to India as the virus spread there.

In spring 2021, as most regions in the world were seeing a welcome downturn in COVID-19 cases, South Asia, driven by India, began what *The Economist* called a "Himalayan ascent." On April 30, the daily COVID count passed four hundred thousand. Nearly one-quarter of the population had tested positive. Deaths were rising, and a public health crisis in one of the world's most populous nations meant a health crisis everywhere. Our vaccine was not approved in India despite the fact that in August 2020 we had submitted the same application to Indian regulators that we had submitted to all other countries in the world. India is very protective of its generic drug industry and makes the registration of vaccines that are manufactured outside India very difficult. We started discussions with the Indian government to see how we could help by providing other medicines that could be used by their public hospitals. On the eve of Pfizer's first quarterly earnings report, as the death toll in India was mounting, I sent a memo to our colleagues in India promising that we would donate to India the Pfizer medicines that the government of India had identified as part of its COVID treatment protocol. We pledged to donate enough to ensure that every COVID-19 patient in every public hospital across the country could have access to the Pfizer medicines they needed in the next ninety days, free of charge. I concluded that Pfizer would stand in solidarity with all those currently affected by COVID-19 in India and around the world and would continue to do everything possible to help.

"As we work to meet the public health need and to be a partner with the Government of India to establish a path forward for our vaccine, please know you and your loved ones are foremost in our thoughts and prayers," I wrote in a letter to our Indian colleagues on May 3.

As the situation in India was deteriorating, the country decided to ban all exports of the AstraZeneca COVID-19 vaccine that was produced locally by the Serum Institute. Unfortunately, COVAX had based most of their planning on this vaccine. This plan was now collapsing, so COVAX returned to us and asked for significantly more doses. We agreed but with one condition: that all our doses go to COVAX's Advance Market Commitment countries, the ninety-two lowest-income countries and economies in the world. This became a sticking point again because COVAX continued its desire to serve all its members. They wouldn't agree to take doses for the poor countries only. They wanted to distribute the doses we were going to give them to all their countries, rich and poor, based on a formula they had agreed to. This was not acceptable to us. Nevertheless, we resumed discussions with a mutual desire to get vaccines where they were needed most.

In the meantime, the US Trade Representative, Katherine Tai, announced suddenly that the US would support a waiver of the Trade-Related Aspects of Intellectual Property Rights (TRIPS) agreement for COVID-19 vaccines at the World Trade Organization (WTO). The TRIPS agreement is an international legal agreement that was signed in April of 1994 and became effective on January 1, 1995. The agreement introduced intellectual property law into the multilateral trading system for the first time. India and South Africa had requested such a waiver from the WTO, but there was no traction for this request so far. Interestingly enough, both countries are among the largest generic medicine manufacturers in the world, and for many people the request looked quite self-serving. One week before the TRIPS waiver announcement, Ambassador Tai had requested a call with me to discuss the global supply of the COVID-19 vaccine. I had the opportunity to explain to her that in the last few months we had invested more than $2 billion for the COVID-19 vaccine, of which more than $1 billion went to manufacturing infrastructure. As a result, we expected to manufacture by the end of 2021 three billion doses of COVID-19 vaccine, of which at least one billion doses would go to middle- and low-income countries. I also reminded her that we had a tiered pricing model and that the lower-middle-income and low-income countries of the world would receive it at cost. I explained that intellectual property is not an obstacle to produce more and that the reasons we cannot manufacture even more doses faster is not the lack of enough manufacturing infrastructure but the lack of enough raw materials. Waiving intellectual property protections could only disrupt the supply chain and make it dangerous for all. I also reminded her that a good way to increase the quantities that go to the poorer countries would be for the US government to allow us to export doses produced under DPA to Latin America and Africa, because right now they did not, and we were exporting to these countries only from our manufacturing sites in Europe. I repeated this specific request a few times, but she didn't respond to it. The call was generally very polite, but when we disconnected, I did not have a very good feeling. During this pandemic, I'd had plenty of discussions with governmental officials, and not all of them were pleasant. However, this was the first time I felt that the person at the other end of the line (or in this case the computer screen) was not really interested in understanding the information I was providing. It was as if she had already made up her mind and the call was made to check a box that said we'd had this discussion. Nevertheless, I never expected that one week later she would announce the US support for India and South Africa's proposal.

When on the morning of May 5, 2021, I read the statement of Ambassador Tai—that the US would support the TRIPS waiver, which would undercut intellectual property protections—I was, frankly, pissed. I was truly angry that at the same time that the US government was blocking us from exporting COVID-19 doses to other countries, their trade representative was supporting a proposal that would directly harm the intellectual property rights of an American company. The US Trade Representative is there to support the interests of the US industry. Instead, I felt the government had just thrown us under the bus. I went so far as to text one of my

White House contacts that I felt "betrayed." Two days later, on May 7, I released a letter to Pfizer colleagues explaining the situation with the global supply. The letter was my attempt to tell things the way they were without sugarcoating anything.

> Dear Colleagues,
>
> The recent announcement that the United States Trade Representative will discuss options to waive some COVID-19 vaccine intellectual property (IP) rights has created some confusion to the world. Has Pfizer done enough to ensure fair and equitable distribution of our COVID-19 vaccine? Is the proposed waiver going to bring solutions or create more problems? I am writing to you today to discuss these questions.
>
> Fair and equitable distribution was our North Star from day one. In order to ensure that every country can have access to our COVID-19 vaccine two conditions had to be met: a price that anyone can afford and reliable manufacturing of enough vaccine for all.
>
> The first condition was met in the early days. Back in June of 2020 we decided to offer our vaccine through tiered pricing. The wealthier nations would have to pay in the range of about the cost of a takeaway meal and would offer it to their citizens for free. The middle-income countries were offered doses at roughly half that price and the low-income countries were offered doses at cost. Many of the poorest communities will receive their doses through donation. Equity doesn't mean we give everyone the same. Equity means we give more to those that need more.
>
> Meeting the second condition was much more challenging but we are getting there with remarkable speed. Thanks to the ingenuity and hard work of our scientists, engineers and

skilled workers, and multibillion dollars of Pfizer investment, we announced that we will provide to the world more than 2.5 billion doses in 2021. In fact, our internal target is 3 billion doses, so we feel quite comfortable about our commitment. Achieving 3 billion doses this year means, by extrapolation, 4 billion doses in 2022. These doses are not for the rich or poor, not for the north or south. These are doses for ALL. We have concluded agreements to supply 116 countries and we are currently in advanced negotiations with many more for a total of approximately 2.7 billion doses in 2021. Upon finalization of all agreements, we expect that 40% of them, or more than 1 billion doses, will go to middle- and low-income countries in 2021.

This clearly poses another question. Until today, we have shipped approximately 450 million doses and the balance is more favorable to high income countries. Why is that? When we developed our tiered pricing policy, we reached out to all nations asking them to place orders so we could allocate doses for them. In reality, the high-income countries reserved most of the doses. I became personally concerned with that and I reached out to many heads of middle/low-income countries by letter, phone and even text to urge them to reserve doses because the supply was limited. However, most of them decided to place orders with other vaccine makers either because mRNA technology was untested at that time or because they were offered local production options. Some didn't even approve our vaccine. Unfortunately, other vaccine producers were not able to meet their supply commitments for varying technical reasons. Most of the countries that did not choose us initially, came back and thanks to our phenomenal supply ramp up, we have started signing supply agreements with them. We expect the supply balance to weigh in their favor in the second half of 2021, and to have virtually enough supply for all in 2022.

Last week, I had the opportunity to provide these facts to the US Trade Representative and explain why the suggested waiver of IP rights could only derail this progress. Which brings me to the second question. Is the proposed waiver going to improve the supply situation or create more problems? And my answer is categorically the latter.

When we created our vaccine there was no manufacturing production of any mRNA vaccine or medicine anywhere in the world. We had to create manufacturing infrastructure from scratch. With 172 years of quality manufacturing tradition, substantial deployment of capital, and more importantly, an army of highly skilled scientists, engineers and manufacturing workers, we developed in record time the most efficient manufacturing machine of a life-saving vaccine that the world has ever seen. Currently, infrastructure is not the bottleneck for us manufacturing faster. The restriction is the scarcity of highly specialized raw materials needed to produce our vaccine. These 280 different materials or components are produced by many suppliers in 19 different countries. Many of them needed our substantial support (technical and financial) to ramp up their production. Right now, virtually every single gram of raw material produced is shipped immediately into our manufacturing facilities and is converted immediately and reliably to vaccines that are shipped immediately around the world (91 countries to date.) The proposed waiver for COVID-19 vaccines, threatens to disrupt the flow of raw materials. It will unleash a scramble for the critical inputs we require in order to make a safe and effective vaccine. Entities with little or no experience in manufacturing vaccines are likely to chase the very raw materials we require to scale our production, putting the safety and security of all at risk.

And I would like to make a final point. I worry that waiving of patent protection will disincentivize anyone else from taking a big risk. We deployed $2 billion before we knew whether we could successfully develop a vaccine because we understood what was at stake. Just recently, I authorized spending an additional $600 million on COVID-19 research and development that will bring our total spend for R&D in 2021 to more than $10 billion. The recent rhetoric will not discourage us from continuing investing in science. But I am not sure if the same is true for the thousands of small biotech innovators that are totally dependent on accessing capital from investors who invest only on the premise that their intellectual property will be protected.

Ending the pandemic and vaccinating the world is a massive, but achievable undertaking. We remain fully focused on getting high-quality, safe and effective vaccines to patients all over the world as quickly as possible and to putting an end to this deadly pandemic. Once again, we will not let politics stand in our way and we will continue doing what we do best—creating breakthroughs that change patients' lives.

Soon after the announcement of this decision, I realized that I was not the only one who was shocked by it. I started having calls with the presidents and prime ministers of many countries, and they were all telling me how surprised they were by this proposal. Off the record, they were all telling me that this proposal did not make any sense. Many went the extra step to tell me that in their view, without intellectual property protections, the world wouldn't have a COVID-19 vaccine right now. Some asked me if I had any idea what the motivation for such a decision was. I told them that in my opinion they didn't expect this waiver to go anywhere, and they just wanted to give something toothless to the progressive wing of the party that was against the pharmaceutical industry. I still remember the response of an EU country leader.

"But it's so irresponsible."

One of the leaders I spoke with was President Ursula von der Leyen. During this call I told her that at least one billion vaccine doses were earmarked for middle- and low-income countries in 2021.

She was a bit surprised to learn that we expected to provide the doses this year and questioned whether this was widely known. To help spread the word, she invited me to speak at a global health conference in Rome. Of course, she also let me know how nice it would be if Pfizer could make the commitments not only for 2021 but also for 2022.

I accepted immediately. In late May, I participated virtually in the Global Health Summit in Rome and shared that Pfizer expected to provide one billion doses of our vaccine to low- and middle-income countries in 2021. And we pledged to deliver an additional one billion doses to these countries in 2022.

"It is our hope that this will accelerate our ability to help save even more lives across the globe. Ending the pandemic and vaccinating the world is a massive but achievable undertaking," I said in my brief remarks.

In the weeks that followed, I calmed down. Rethought things. Despite my disappointment with the US Trade Representative's decision, having shadows over our relations with the administration was not a good feeling. And, more important, I believed there were many meaningful things we could accomplish together. I wouldn't let my anger or emotion get the better of me. I felt the best way to forge ties with this White House was to show our true character and propose solutions to the enormous health challenge that COVID-19 was still posing. The US was feeling the heat for not allowing exports of American-produced doses or having a plan to share the vaccine wealth with the less fortunate. I knew we could help.

As part of my desire to move on, I arranged a call with the White House COVID-19 response coordinator, Jeff Zients. I had spoken to Jeff many times previously for issues related to the US government response to the COVID-19 crisis. Since January, when he took over his role as a member of the Biden administration, he had led the US response very successfully. I respected him as someone equally at home in government and business, someone I had confidence in and with whom we could work. However, since the incident with the TRIPS waiver announcement, I had not called him. Apparently, Jeff had heard about my ire, and he hadn't called me either.

The call began a bit tentatively. Then, to my surprise, Jeff shifted gears and said first that we should move forward, and the best way to do that was to find something big that we could do together. I grabbed the opportunity and offered Jeff an idea: "How about if we work together and we produce hundreds of millions of doses and make them available to the United States government at cost? Then the US government can donate them very quickly to the neediest countries."

Jeff liked the idea, and we started brainstorming about it. We discussed which countries we could give the doses to and agreed that the ninety-two countries of Gavi and the countries of the African Union could become an excellent recipient pool. Then we discussed the logistical challenges of the US government sending doses to so many countries, and I promised that we would take care of it and ensure the complex logistics in getting these vaccines to countries with limited infrastructure. The idea was shaping into something big, a Biden vaccine Marshall Plan.

Jeff did not hesitate. He was enthusiastic and eager to take this proposal to the president. We went to work, starting legal and logistical discussions. The US government wanted to use the COVAX facility as a main mechanism for distributing the doses, but we were insistent that if the doses were donated by the US government to the COVAX facility, the doses must only be made available to the ninety-two Advance Market Commitment countries and the African Union, where the needs were greatest. The US made this a condition to COVAX and decided to purchase five hundred million doses with an option to procure more. It was a major step toward vaccine equity, and I was proud that we were leading the way. There was, however, one stubborn issue that needed to be resolved first with respect to liability-related provisions. In the US, the Public Readiness and Emergency Preparedness (PREP) Act provides broad protections to biopharmaceutical companies and others that make medicines and equipment used to combat a pandemic against certain liabilities during the public health emergency. It also provides compensation to individuals if they suffer harm from these essential products. But most of these countries do not have these protections. And in a few instances, especially in low- and middle-income countries, the governments or judicial systems are not well equipped to adjudicate claims against manufacturers, particularly in the context of a pandemic. We therefore had to discuss with governments solutions that were fair to both sides and would ensure that the company had appropriate liability protections given the unprecedented circumstances we were facing. To be clear, we have always had confidence that our vaccine is both safe and effective, but because of the sheer number of patients receiving our vaccine, individuals in countries around the world could blame our vaccine for health conditions that are completely unrelated to the vaccine, bring claims against us in numerous different courts, and put us at substantial financial risk.

While I was concerned about the potential for litigation, however unfounded, or requiring my team to navigate foreign legal systems, I also knew that we must find a solution, and not allow the challenge of liability protections to be a barrier for lifesaving vaccines to reach the poorest countries in the world.

After weighing the pros and cons, I decided to move ahead with this option and honor our commitment to equity. By agreeing to route the five hundred million vaccines through the COVAX facility, we were able to obtain legal protections for our company, even if less than what is offered under the PREP Act, but most important, it allowed us to ensure that our doses would be provided to countries that were so desperately in need. I knew I was taking a risk by agreeing to deliver doses to these countries in this manner, but I felt it was a reasonable option for the company, and I could sleep better at night.

Once the details had been sorted, Jeff reported that President Biden wanted to announce this arrangement while he was at the G7 meeting scheduled during the second week of June in Cornwall, England. I immediately recognized the genius of the president's idea. With the media all focused on this summit, this would be a great opportunity to demonstrate that America was stepping up again to do the right thing for the world and restore the country as a generous diplomatic partner on the world stage.

As the G7 gathering was only a few days away, we scrambled to make our travel arrangements. Because of the lingering COVID restrictions in Europe, we needed a special waiver from the UK government to enter the country. My team literally worked around the clock and, with the help of the White House, managed to obtain this approval only hours before our scheduled takeoff time.

On June 9, during the flight over, I read a *New York Times* story with the headline "Biden Aims to Bolster U.S. Alliances in Europe."

"Biden, who will arrive for a series of summit meetings buoyed by a successful vaccination program and rebounding economy, will spend the next week making the case that America is back and ready to lead the west. Mr. Biden's overarching task is to deliver the diplomatic serenity that eluded such gatherings during the four years in which Mr. Trump scorched long-standing relationships with close allies, threatened to pull out of NATO and embraced Putin and other autocrats, admiring their strength," the article read.

As I stared out the plane window, a smile came to my face with the thought that tomorrow the world would be surprised with this major announcement. We would create waves around the globe. I had little time to enjoy that good feeling because the news feed on my cell phone began to fill with headlines, first from *The New York Times*, then *The Washington Post*, *The Wall Street Journal*, and others. The story of our vaccine plan had leaked less than twenty-four hours before our press conference. Representatives from Prime Minister Boris Johnson's office reached out to learn more details. We were obligated to maintain the embargo on the news and gently suggested they reach out to President Biden's office for information. My team and I worried that, now that the news was out, the Biden team might cancel the event, but we kept moving forward with our plans. I read my remarks one more time to prepare.

That evening, we arrived at Cornwall Airport Newquay and drove a short distance to a beach hotel on the coast. Sally Susman and our security detail, Tim Conway, were with me. Tim is not only the person responsible for ensuring executive protection but also a focused, detail-oriented man who is invaluable in solving complex logistical problems. Tim, having come to us after a twenty-four-year career with the US Secret Service, was working his White House contacts and global security network to ensure we had the access, waivers, and credentials we needed. Tim is a pleasure to be with, and though he's the epitome of discretion, we coaxed him to tell us a few stories from his days in the Secret Service guarding various presidential families. These quiet times, in remote places with my team, are special moments for me. We get to know one another, laugh a little. It felt good to be back on the road again with my colleagues.

The next morning, a typically foggy English day, we worked in our hotel rooms until it was time to travel along the narrow, shrub-lined country roads to the sprawling seventy-two-acre grounds of Tregenna Castle, the G7 meeting venue. After a long walk through multiple security checkpoints, Sally and I met up with President Biden for a quick chat and a review of the logistics. The setting was spectacular: the presidential podium standing on a green lawn with magnificent trees in the background.

The president strode out first, and I followed closely behind. I then watched him make his remarks and listened with deep pride as he made the very particular point that these vaccines "are mRNA." I took that to mean that the US was shipping the good stuff.

In my remarks, I applauded President Biden's leadership, noting, "As the G7 countries come together for this critical summit, the eyes of the world are on the leaders of these powerful nations to help solve the ongoing COVID-19 crisis facing the neediest of our global neighbors. While great progress has been made in many developed nations, the world is now asking the G7 leaders to shoulder the responsibility to help vaccinate people in all countries."

I continued: "Mr. President, I know from our conversations that we agree that every man, woman, and child on the planet—regardless of financial condition, race, religion or geography—deserves access to lifesaving COVID-19 vaccines." And then, as the moment made clear, "Once again, the United States has answered the call."

As I finished and the president and I were walking away from the bank of cameras, I was surprised and delighted by a hug from First

Lady Dr. Jill Biden. In the past I'd had the opportunity to talk to her through a Zoom call, but I had never met her before in person. She was so warm and exuded a steady calm and clear focus. Dr. Biden, the president, Sally, Jeff Zients, Secretary of State Antony Blinken, and I chatted for a while, enjoying a light conversation in the bucolic setting. Someone noted that so many in America's fight against COVID have origins as immigrants to our country. I squared my shoulders at the thought of being one of them.

As we started to say our farewells, the president pulled something from his pocket and passed it to me in a handshake.

"This is what commanders on the battlefield give to someone who has displayed extraordinary courage. You deserve this," he said.

It was a heavy metal coin with the presidential seal on one side and the name JOSEPH R. BIDEN, JR. engraved alongside the state of Delaware on the other. I felt tears in my eyes. I tucked it into my wallet for safekeeping. I'll treasure it as a true token of our shared hopes for what these powerful vaccines can do for the needy people who will receive them.

9

Navigating a Political Minefield

"Man is by nature a political animal."

—Aristotle, 384–322 BC

VACCINE EQUITY WAS ONE OF our principles from the start. Vaccine diplomacy, the idea of using vaccines as a bargaining chip, was not and never has been. But, like an embassy, my office seemed to be on speed dial for world leaders.

The history of vaccine diplomacy is rooted in the early wars between France and England. The British doctor Edward Jenner published research on the use of cowpox (vaccinia) virus in the prevention of smallpox in humans. In the 1800s, smallpox vaccine was shipped across the channel to France. Napoleon caught on to its importance in foreign relations and welcomed Jenner to the National Institute of France. Napoleon decreed that vaccine departments be built across France. During those French-English wars, diplomats negotiated prisoner exchanges and other terms. Napoleon saw vaccines as a bargaining chip and wanted to use them in the negotiations. But Jenner disagreed, famously writing to his colleagues that the "sciences are never at war."

During the COVID pandemic, almost every head of state reached out to thank me, to ask for more supply, or to complain that they were not getting more. Often my well-intentioned staff offered to triage the calls or return them for me. While I appreciated their intentions, I knew it was important that I accept and personally take every call we received from heads of state (presidents, kings, prime ministers, and tribal chiefs) or ministers of health. They represent their citizens, and we serve their citizens as our patients. Without exception, the leaders were eager to hear news about the vaccine and to get in line for what was initially a limited supply.

As with most families during the pandemic, my son, Mois, and my wife, Myriam, shared dinner responsibilities. Mois relished being part of the household action. He loved to prepare communal meals with expertly cooked steaks and pasta, chicken curry, and other set menus.

Over dinner, his colorful commentary after listening in on some of these calls made me laugh.

"That guy is not very smart," he'd observe. "The guy yesterday was really sharp."

"She was not interested at all in what you had to say. She just wanted to check the box that she talked to you."

"They seemed uninterested in antigens, but they sure latched on to the issues about temperature and cold storage."

"I can't believe how fast those negotiations happen."

Political negotiations were tense, and so I was over the moon to have Mois with me. When he was younger, we once sat together and listened to my father's brother, my uncle Into, speak on a recorded oral history about his brother, Mois's namesake, because I wanted both of us to understand their lives and stories. In between cell phone and video calls, I would often discuss with Mois what I was going to say, sometimes just bounce it off him so that I could hear it out loud. He happily provided that service.

My daughter, Selise, was home for part of the pandemic but was eager to get back to her friends in the dorm. Like me, she is funny and has an opinion about everything. We also share similar ambitions and drive. We both want to help people. She is incredibly generous.

While living back at home like so many of her classmates during the pandemic, Selise sensed the buzz and tension that filled our house. She was sensitive to the energy in the house, and sometimes had to be cajoled to come downstairs to dinner. Mois would yell, "I am going to eat all the French fries!" This would induce movement upstairs, and she would join us at the dinner table. Once there, she would quickly engage in whatever discussion was underway. Both kids seemed to appreciate that something important for society was happening. Too often I would have to leave the table to take a call.

Too often, according to them, I would also feel the need to tell a dad joke, also known as a bad joke.

As we entered the fall of 2020, the days leading up to and just after our announcement of the vaccine's efficacy, the phone calls to me escalated from the bureaucrats to the ambassadors to the heads of state. Global cases of COVID-19 approached fifty million. I spoke frequently with political leaders over the phone or via videoconference. One leader insisted that we meet in person—the prime minister of Albania, a country that neighbors my home country. He expressed concern that they might not receive doses since they were not members of the European Union. He reminded me of the high flow of people between Greece and Albania. I was so touched by his determination to help his people. His focus was on frontline workers. With masks and proper distancing, we managed to meet in New York City. Later, when he returned home, he sent me a beautiful painting he had created.

A few days later Jonathan Nez, president of the Navajo Nation, a sovereign nation inside the United States, spoke to me about a second surge they were experiencing. He said the tribe's Institutional Review Board (IRB) appreciated our approach to informed consent, and his advisory committee preferred to work with Pfizer. "The vaccine can't get here soon enough," he said. I told him that the US government would control the distribution of our vaccine. The president said that he was also worried about vaccine hesitancy among tribal members. He had a videoconference scheduled for his people. He wanted me to join but worried that my schedule was too busy. I said, "If you think one hour of my time can save a life, I am happy to join." And I did.

The Japanese head of state called me during a visit to the United States. He was eager to discuss doses for his people, which we were able to provide. Once we concluded, I told him that we would like to help further, and I offered our assistance for the upcoming Olympic Games to be hosted in Japan. Athletes and delegates would be in close proximity to one another. I thought it would be a terrible loss if the Olympics had to be postponed for a second time.

For me, the Olympic Games are a great demonstration of modern democracy. In the past, wars and conflicts were paused during the Olympics. The modern games got their start in 1896 after the French baron Pierre de Coubertin became so inspired by the Olympic ideal that he formed the International Olympic Committee (IOC). Today the Olympics brings people from around the world together in a celebration of our connected humanity. I was seriously concerned that canceling the Olympics would signal that a virus had won, had beaten our global civilization.

After speaking with the Japanese prime minister, I called the IOC and offered our help. I told the IOC president, Thomas Bach, that we wanted to help vaccinate athletes and their delegations. The number of doses required was not large. The logistics, however, were more significant. In May 2021, we announced the signing of a Memorandum of Understanding with the IOC to donate doses of our COVID-19 vaccine to help vaccinate athletes and their delegations participating in the Tokyo 2020 Olympic and Paralympic Games to be held in late summer 2021. "This donation of the vaccine is another tool in our toolbox of measures to help make the Olympic and Paralympic Games Tokyo 2020 safe and secure for all participants and to show solidarity with our gracious Japanese hosts," said President Bach.

It was an important, symbolic victory for science and humanity. Later, President Bach not only extended an invitation to me to attend the opening ceremonies of the Olympic Games but, to my surprise, also invited me to carry the torch. I was both honored and humbled. Unfortunately, due to increased COVID restrictions, I was ultimately unable to carry the torch, but I did attend the opening ceremony, and the experience was among the highlights of my life. And President Bach was kind enough to give me a torch, which I proudly brought home with me.

Occasionally, enlightened leaders would look beyond the current crisis to express interest in the science. How can our brightest scientists collaborate with your best scientists to prepare for the next crisis? The UK's prime minister, Boris Johnson, was one of those leaders. Unfortunately, during our discussions, we nearly created an international incident. The prime minister and I spoke on January 14, 2021, in the early evening, London time. Being the first country to approve our vaccine, the UK had already vaccinated 2.5 million people and had 40 million doses on order. He wanted to increase that number. "We are losing the elderly," he told us. They had the people available now, and the healthcare system to make it happen. They planned to move three million doses from February 2021 forward to January 2021. If Pfizer had more doses, the UK was keen to use them. From my perspective, the UK was doing an exceptional job under tremendous pressure. It also had an emerging variant, originating from India, that was creating new concerns and months later would be known as the delta variant. Mikael Dolsten spoke up to assure the prime minister we were following the variant carefully and believed our vaccine would protect against it. On the basis of the knowledge we had, I committed that we could meet his request for more doses. By the end of the call, a photograph was taken of the videoconference screen that the prime minister later posted to LinkedIn, and he told us he would ask the UK to applaud Pfizer during one of its evening rallies. My team and I left that meeting full of pride.

Later in the day, however, I learned that our team did not believe we could deliver on the promised supply. A member of the team had made an honest mistake and double counted the available doses, and we didn't actually have three million doses unless we took them from someone else, which of course would not be fair or ethical. I felt awful and had to call the prime minister back. He was gracious but told us the news hit very hard; they were depending on the doses to meet goals. I told him we would work through the weekend to explore every possibility to increase the scheduled doses. I had made a personal commitment, and because of this the UK prime minister had made a promise to the people of his country. I was not going to take that back. I was honor bound.

Our team went back to the drawing board. We identified a surplus of doses—created by the slowness of the US government to distribute doses in January, during the last weeks of the Trump administration. However, vaccine doses manufactured in the US using the DPA could not be exported internationally. Fortunately, those surplus doses had been manufactured at our facility in Europe and sent to the US. There was no prohibition on exporting doses that had not been produced in the US. At that time, the UK was the only country vaccinating so quickly that demand surpassed supply. As a result, we worked on a plan to meet the UK's needs, examined it carefully, and gave the green light to ship those doses to the UK.

We scheduled a call with the prime minister for our morning time on Martin Luther King Day, Monday, January 18, 2021. I told him we had good news. I took personal responsibility and apologized for the concern we had caused. The prime minister appreciated the transparency and the effort. He thanked us all and then added that he had studied up on mRNA. He told me he felt it was one of the more important technologies of the future. It was something he wanted to work with Pfizer on at every level to help protect our populations in the future so we could make sure when new variants arise, we could rapidly turn things around. "I hope that our teams can take that forward."

For my own team, what I hoped to model here was a set of values that mean a lot to me. Take responsibility. Be accountable. Don't try to hide. Figure out how to fix it and apologize directly. Explain what you have learned.

Like the UK, Canada had low vaccine hesitancy and pushed for more doses sooner. The Canadian prime minister, Justin Trudeau, was concerned that his country was getting in line behind European countries, and I assured him that Canada was well positioned, but also acknowledged that negotiations with Europe were important to our mutual interests since our site in Puurs, Belgium, was a critical part of the solution and we could not afford upsetting Europe and risk shutting down exports. I'd already heard from the Trump White House that it wanted all US supply for US purposes. Canada would get more than it expected, just not in the next few weeks.

Over the following days I had productive calls with the president of Mexico and others in Latin America. All were facing the dual challenge of a health crisis and an economic crisis. They felt they had to solve for health first. Our call with Brazilian officials, including President Bolsonaro, was heartbreaking. The country had been slow to order doses and had initially preferred Chinese and Russian vaccines. The slowness and miscues created political problems for them, and worse, their people were dying.

On January 20, 2021, in his first act as the new president, Joe Biden paused his inaugural address to pray for those who'd died during the pandemic. "We must set aside the politics and finally face this pandemic as one nation," he told a global audience.

In April, as more and more people were vaccinated, I flew to our manufacturing site in Puurs to tour our facility with the European Commission president, Ursula von der Leyen, and the prime minister of Belgium, Alexander De Croo. It was the first time I had the pleasure of meeting, in person, Özlem Türeci, of BioNTech. A brilliant scientist, she'd cofounded the company with her husband, Ugur, and I quickly learned that we share very similar personalities. My team was impressed with not only her eloquence but also her warm personality.

During our visit, we were introduced to a team of colleagues who stood in front of a large visual timeline of our work together. It was really very emotional, and I offered my heartfelt thanks. Özlem spoke about how BioNTech was so grateful to have a partner with whom they worked so well and who could bring their discoveries to life. I said that the Pfizer-BioNTech partnership was "not only about science, not only about competencies, but also about the shared value to serve patients." Later in the day, Özlem and I explained to President von der Leyen and to Prime Minister De Croo our approach to vaccine candidate selection, prioritizing safety data and immunogenicity. Our method was proving valuable with the various variants that were arising in the UK, South Africa, Brazil, and elsewhere. My message was simple: the power of science, the need for a vibrant private sector, and collaboration at all levels, including with governments.

Standing in our manufacturing facility in Puurs, President von der Leyen noted that this place is a symbol for the European Union's fairness and openness. Europe, she noted, was producing vaccines for Europeans and for citizens around the world. The EU had exported more than 155 million doses of vaccines to over eighty-seven countries worldwide since December.

"We are the pharmacy of the world," she exclaimed. "And we Europeans take pride in this and we invite others to join. Because we all know: Nobody will be safe until everybody is safe."

She thanked the EU's strong and reliable suppliers, like BioNTech-

Site lead Luc Van Steenwinkel, Pfizer CEO Albert Bourla, BioNTech cofounder Dr. Özlem Türeci, European Commission President Ursula von der Leyen, and Belgian Prime Minister Alexander De Croo during a tour of Pfizer's Puurs, Belgium, COVID-19 vaccine manufacturing plant, April 23, 2021.
Photo courtesy of Pfizer Inc.

Pfizer. "Indeed, the main vaccine used so far in the European Union is the one produced right here in Puurs, in Belgium—a true vaccine powerhouse. And with the enormous efforts of BioNTech-Pfizer and the acceleration of their vaccine deliveries, I am now confident that we will have sufficient doses to vaccinate 70 percent of the adult population in the European Union already in July. And this pioneering technology we see here will help us with that."

The president noted in her speech that Pfizer and BioNTech had delivered consistently, and for that reason the European Commission would sign the world's biggest vaccine supply deal ever—up to 1.8 billion doses.

The story of our approach for Israel was very different and merits a chapter of its own.

10

A Beacon of Hope

"What is common to all men? Hope. Because those who have nothing else possess hope still."

—Thales of Miletus, 620–546 BC

WHEN WE STARTED THE FIRST vaccine shipments to people around the world, the excitement was very high. But then the reality hit. I had been telling everyone that the vaccine could come by October 2020, but it turned out that very few people besides me were expecting this really to happen. Clearly, the delivery machine was not ready. In almost every country, the ability of the health authorities to administer the available doses was not stellar. In the beginning, there were more doses available than what they could absorb. And when the vaccination rates started picking up, then the production could not meet the demand and became a bottleneck. I realized that it would take a long time to have enough people vaccinated so that we could see tangible, measurable results. The hope created by the first vaccinations faded quickly as daily life remained the same. We needed an example to demonstrate to the world that the hope was real. The idea was to select a country that, if it was given uninterrupted supply, could ensure quick vaccinations all the way to herd immunity and demonstrate the impact on health and economic indices. Ideally, the country would be relatively small and have high standards for its health system and good electronic medical records.

Benjamin (Bibi) Netanyahu, the prime minister of Israel at the time, was among the first to call me to advocate for earlier deliveries in his country. Our initial conversation was very casual. He told me he had found out I was a Greek Jew from the Greek prime minister, Kyriakos Mitsotakis, who was very proud of me. As we continued our discussion, I revealed that Mikael Dolsten, our head of research, was a Swedish Jew who during medical school obtained a fellowship at the Weizmann Institute, a renowned Israeli scientific institution.

"Now I am really pissed with our intelligence guys," the prime minister responded, his amusement barely disguised. "They didn't tell me this either."

I was thinking that it wouldn't make a difference for the discussion we were about to have, but I told him jokingly to let it go because I didn't wish to be between Mossad and his frustration with them. After these initial talks he started pushing for earlier deliveries to Israel. As with all other heads of state, I explained the supply situation, asked for his patience, and promised to do my best. Netanyahu knew a lot of technical details about the virus and the different vaccines, and he was the first leader who spoke about herd immunity. His knowledge and direct style caught my attention and started me wondering if Israel could be the country to demonstrate the benefits of widespread vaccination. I called Dr. Luis Jodar, our tenacious Chief Medical Officer for Vaccines, who has a strong commitment to patients. I asked Luis the question, and he was enthusiastic. Israel had the right size, with tightly controlled borders; it had an excellent healthcare system, with electronic medical records for more than 98 percent of its citizens; and the country was experiencing a very tough situation with COVID-19 infections. Healthcare in Israel is universal, with government-funded participation in one of four nationwide medical insurance programs that operate as health maintenance organizations. All Israeli residents are assigned unique identification numbers that enable data linkage in the national medical records database. I asked Luis to include Israel on the list of candidate countries, and to calculate the number of doses it would require for herd immunity.

Two days later, I received another call on my cell phone from Prime Minister Netanyahu. I was surprised because it seemed too soon for a second call, and he hadn't even bothered to schedule it. He'd just called. He brought to my attention some legal issues related to our contract negotiations for delivery of the vaccine. I was not aware of the details and promised to come back to him. Doug Lankler, our General Counsel and a trusted advisor, explained to me his concerns, and I called Netanyahu back. A few hours later, he returned my call. Ron Dermer, Israel's ambassador to the US, was already on the line, together with Israel's national security advisor. I looked at my watch and calculated the time in Israel. I was shocked. "Prime minister, it is two thirty in the morning!"

"Don't worry about that. I don't need much sleep," he replied. "Look, Albert. If we leave this just to the lawyers, we will never get it done. I will conference our head lawyer now. Can you do the same?"

I felt a little bit uncomfortable, but I was also intrigued by his decisiveness and sense of urgency. I conferenced Doug, and Netanyahu conferenced a couple of lawyers. I think they were from a US legal firm representing the state of Israel. In the discussion that followed, I was impressed by Ambassador Dermer, who came up with some brilliant solutions. By 3:00 a.m. Israel time, most of the issues were resolved, with only a few of them remaining on the table. For those, we agreed that the legal teams would explore solutions during our morning. The initial contract was signed a couple days later, and I thought this was the last time I would hear from Bibi.

In the meantime, Luis had calculated the doses we would need for Israel to reach herd immunity, and I kept staring at the list of countries fitting the criteria for an observational study. Which one could execute faster and better? Israel was at the top of the list, but there was a problem. If that country was selected, everyone would think it was because of my Jewish heritage. I was thinking about engaging our office in Israel to answer some of my questions about their vaccination plans, but suddenly the phone rang again, and Netanyahu was on the line. Ron Dermer was with him. The prime minister started talking again about the need to receive more doses because of the bad situation in Israel. I explained that unfortunately this was the reality all over the world, but I started asking questions about their vaccination program. Once more I was impressed by his knowledge of the details. He spoke about their hospitals, his intention to deploy the army, his "Green Pass" idea, and their plans to address vaccine hesitation. He was clearly on top of things. Suddenly, Dermer suspected the importance of this to me and started selling even more of Israel's ability to execute in crisis: "We are a society that has learned to live under constant threats. Free spirited as we are during periods of peace, we are the most disciplined during periods of crisis." Netanyahu synced immediately with Ron and started hyping Israel's capabilities. The next day, I spoke to Mikael, Luis, and Doug about the possibility of selecting Israel as the country to demonstrate to the world what our vaccine could do. We all agreed that Israel could be the ideal place to do this. "You know everyone will think we did this because we are Jews," I texted Mikael.

"I know, but they are the right bet," he replied.

I planned to call the prime minister to discuss the possibility of a real-world evidence research collaboration where we would have observational, nonrandomized, noncontrolled data. We would provide doses under the previously negotiated vaccine supply agreement, and in return Israel would commit to moving very fast and publishing the real-world efficacy data. Of course, the next day he called, before I even had a chance to call him. I brought him the idea for a research collaboration. In a little more than a week, we signed a research collaboration agreement with the Ministry of Health, and we formed a steering committee with epidemiologists from Israel, Pfizer, and Harvard University.

In time, the story of Israel would inspire and inform the world.

By early March 2021, we were preparing to publish interim results from the vaccine campaign in Israel. I wanted to be there on March 11 because it would be the one-year anniversary of the World Health Organization proclaiming COVID-19 a pandemic. My plan was to visit countries in the area, including Israel, but several obstacles stood in the way. One of those obstacles was the coming Israeli elections. The legislative elections were hotly contested, and the potential for my visit ahead of the vote was leaked to the press, creating a few local news reports. Because of my family's story of narrow escape from the Holocaust and Pfizer's success with the vaccine, my profile there was high. Many people reached out to me with advice. Half told me not to come because it might interfere with the process. The other half insisted I come.

The other obstacle was my own vaccination. Getting vaccinated had created a crisis of conscience for me. On the one hand, I was eager to show the world that the Pfizer CEO was first in line because he had confidence in the vaccine. On the other hand, I didn't want to step in front of those who needed it more. In those early days of the vaccine, there were long lines awaiting the jab. Vaccine hesitancy was low among those who needed it most. Demand was high, and the states were still figuring out logistics. I decided to wait my turn. Not to cut the line. My predicament, however, was upsetting because I wanted to do what was right. We debated the decision. We looked at opinion results showing that the public wanted to see their own doctor and other healthcare leaders get vaccinated, thus reassuring patients it was safe. But if I was vaccinated, I would want my team to be vaccinated, and so on along the line of management. I chose to wait until my vaccination might be used to encourage those with vaccine hesitancy later on.

Nevertheless, Netanyahu asked me repeatedly during our calls if I had been vaccinated. Israel was moving very quickly, and he said constituents kept asking him if I had taken it. I knew that he also cared deeply about me. During this difficult period in which we were working so closely together to find solutions to a devastating pandemic, we became very close and we shared many personal stories. As we were usually talking during late evenings Israel time, he would talk to me about his difficult day or his family, and I would do the same.

"I know I am like a Jewish mother to you on this, but you must take the vaccine," the prime minister insisted. Then his wife, the First Lady of Israel, leaned on him. "My wife, who is sitting next to me, is asking when you plan to vaccinate yourself. What shall I tell her?" he added. He even called out the army, in a sense. He said the Israeli army always vaccinated commanders first.

In mid-February I eventually received my first dose and then the second dose on schedule twenty-one days later. But as March 11 approached, it became clear that I was not going to meet Israel's requirements for entry. According to Israel's Green Pass rule, travelers are allowed to visit Israel one week after receiving their second dose of the vaccine. I was not going to meet that timeline and, therefore, was ultimately ineligible for entry into the country.

I did not visit Israel. Many people were very disappointed with my decision. Many others were very happy. They thought that my visit could send a political message amid a polarized election period. I hoped it could have helped to send a clear message that any country can overcome the pandemic by prioritizing the right public health decisions, as Israel did. But we were able to achieve this without my going there.

Later, in an article for *The Lancet* (a renowned scientific magazine), several months into the Israeli vaccine campaign, Luis joined others in sharing the results. Using national surveillance data from the first four months of the nationwide vaccination campaign, we analyzed incident cases of laboratory-confirmed COVID-19, as well as vaccine uptake in residents of Israel aged sixteen years and older. Israel has a population of 9.2 million citizens, 6.5 million of whom were vaccine eligible because of their age. The nationwide vaccination campaign initially targeted healthcare workers, long-term care facility residents, immunocompromised persons, and the elderly. Vaccination was subsequently offered to younger age groups, and by February 4, 2021, anyone sixteen or older was eligible for vaccination. Vaccination rates reached more than 220,000 people a day, a rate comparable to 8 million people in the US. Less than three months later, on March 11, 2021, the first anniversary of the pandemic declaration by the WHO, the Israeli Ministry of Health released data showing greater than 97 percent efficacy for symptomatic cases and 94 percent efficacy for asymptomatic cases.

Two days later, the *Financial Times* reported that "Israelis raise glass to Pfizer as lockdown ends." *L'chaim*, Pfizer. Israel had lifted its lockdown, and the world was regaining hope that liberation was coming. By mid-April, more than ten million doses of the vaccine had been administered. Approximately 70 percent of Israelis sixteen and over received two doses. Ninety percent of those sixty-five and over received both doses. The number of cases fell precipitously.

That spring, Passover and Easter news reports from Jerusalem showed that the crowds commemorating Good Friday were enormous, so large in fact that it was as though the pandemic had never happened. A Roman Catholic priest was quoted saying it was like a miracle. Easter 2021 was much brighter than Easter 2020 in the Holy City. The Israel Ministry of Health planned, organized, and continues to lead the nationwide vaccination campaign.

The anecdotal reports were heartwarming, and the epidemiological evidence historic. According to the article in *The Lancet*:

> Israel provides a unique opportunity to observe the nationwide impact on SARS-CoV-2 transmission of a rapidly increasing percentage of the population with vaccine-derived immunity. SARS-CoV-2 transmission is likely to continue until the proportion of the population with immunity exceeds a herd immunity

> threshold, which has been estimated to be at least 60%, although the emergence of more transmissible SARS-CoV-2 variants could result in higher herd immunity thresholds. Achieving the SARS-CoV-2 herd immunity threshold might not be reached, however, without vaccinating some individuals younger than 16 years. In addition, the duration of immunity to SARS-CoV-2, either from infection or immunisation, is not known, and progress towards herd immunity in Israel could be disrupted by the emergence of new SARS-CoV-2 variants if those variants are less susceptible to the current vaccine-induced immune response and if they were to become broadly disseminated. Further studies are needed to monitor the population level of immunity, identify disruption of viral transmission, and detect and evaluate the effects of emerging SARS-CoV-2 variants.

Separately, the Israeli Ministry of Health published a preview in *The Lancet* demonstrating that a country with a population approximately thirty-five times smaller than the US prevented 158,665 infections, 24,597 hospitalizations and 5,533 deaths in its first 112 days with the Pfizer-BioNTech vaccine. That is the equivalent, in just 112 days, of more than 193,000 fewer deaths in a country the size of the United States. That's the population of a midsized American city. I tweeted in mid-May, "While caution should be used in extrapolating to other countries, these observational findings are demonstrating the impact we have in reducing human pain."

Meanwhile, in the United States, the Centers for Disease Control was able to cite the Israeli example to reassure Americans. The CDC reported that data from Israel suggests people vaccinated with Pfizer-BioNTech COVID-19 vaccine who develop COVID-19 have a fourfold lower viral load than unvaccinated people, adding: "This observation may indicate reduced transmissibility, as viral load has

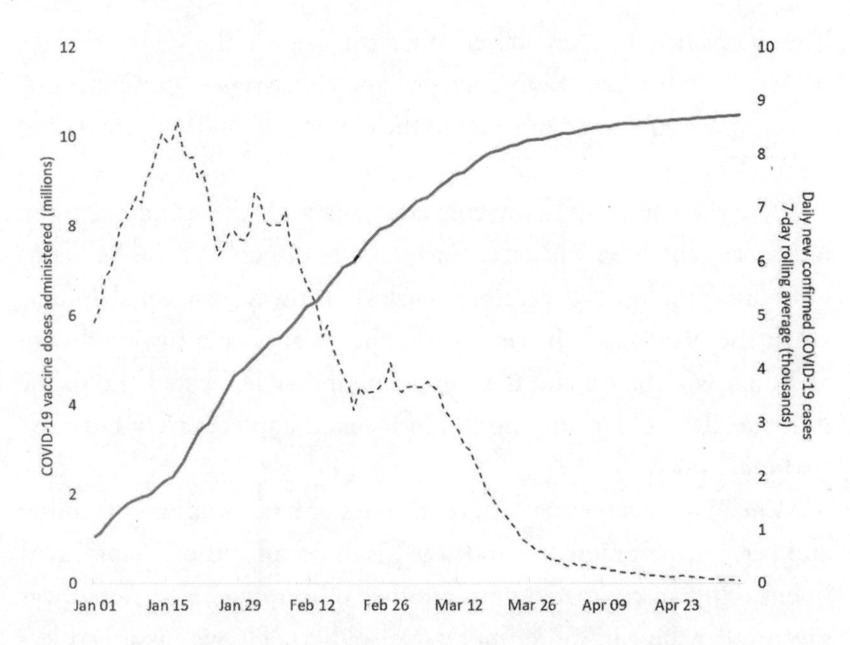

Data from spring 2021 study of Israeli vaccination rates vs. COVID-19 cases

Adapted from Hannah Ritchie, Esteban Ortiz-Ospina, Diana Beltekian, Edouard Mathieu, Joe Hasell, Bobbie Macdonald, Charlie Giattino, Cameron Appel, Lucas Rodés-Guirao, and Max Roser (2020), "Coronavirus Pandemic (COVID-19)." Published online at OurWorldInData.org. Retrieved from: https://ourworldindata.org/coronavirus [Online Resource]; licensed under CC BY 4.0.

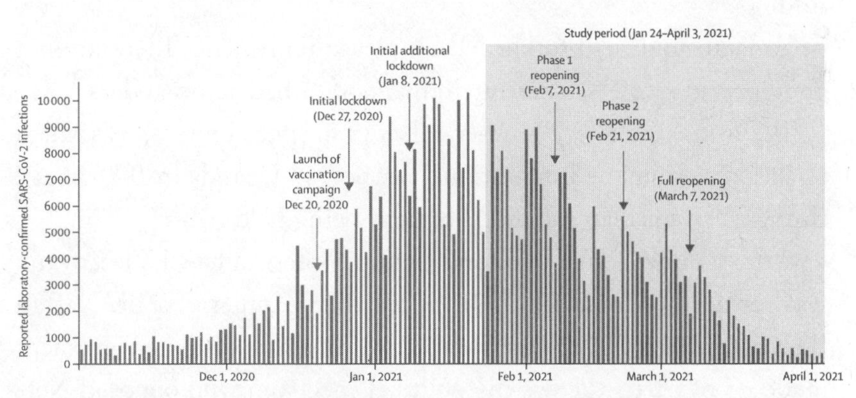

Daily laboratory-confirmed SARS-CoV-2 infections in Israel (November 1, 2020, to April 3, 2021)

Reprinted from E. Haas et al., The Lancet *397, "Impact and effectiveness of mRNA BNT162b2 vaccine against SARS-CoV-2 infections and COVID-19 cases, hospitalisations, and deaths following a nationwide vaccination campaign in Israel: an observational study using national surveillance data," 1819, copyright © 2021, with permission from Elsevier.*

been identified as a key driver of transmission." The CDC also reported, "High Pfizer-BioNTech vaccine effectiveness (92%) against infection was observed in Israel in the context of multiple circulating strains."

Over the course of many conversations with the prime minister, one story remained with me. Early in our discussions, he had congratulated me on the vaccine's success. I always corrected leaders when they said that. "In fact, it was the great people I worked with at Pfizer who had made the impossible possible," I told the prime minister. It was a team effort. He listened appreciatively but then told me a story.

When Netanyahu was in his early days of Israel's military training and service, he graduated and was given command of a small regiment. He had competed with another officer, whom he noted was given taller, bigger, more impressive soldiers. He was assigned less physically impressive soldiers. Later, he complained to his brother, Yonatan, a more senior and more seasoned commander: "Look at what happened to me," he said. "They didn't give me the good soldiers."

Yonatan told his brother, "Bibi, please remember. There are not good or bad soldiers. There are only good or bad commanders."

The prime minister's brother, a leader in special forces, was killed in action at Entebbe International Airport in Uganda in 1976 at age thirty when his team rescued hostages being held there.

After a drawn-out and bitter election contest in Israel, Netanyahu was replaced on June 13, 2021, by his former chief of staff, Naftali Bennett, as prime minister. Bennett formed a coalition government made up of parties across the political spectrum who opposed Netanyahu for a variety of political and other issues. But it became clear from the outset that when it came to defeating the virus, there was virtually no daylight between the new government and the old. I picked up a dialogue with Bennett as I had with Netanyahu, and our scientists and researchers continued working with the Israeli scientists as if there had been no election at all. Even though the top ministers had changed over with the switch in government, there was remarkable continuity between the teams working directly with Pfizer on our important project.

Bennett initiated our personal relationship at the end of June, approximately two weeks after he took over. Israel had a small number of doses that were approaching their expiry date, and he wanted us to facilitate moving these doses to another country so they could be used immediately. The plan was to give these doses to the Palestinian National Authority, but at the last moment the Palestinians refused to accept them, and now there was an urgent need to resolve this. There were a lot of challenges, but I agreed with him that it would be terrible if we let lifesaving doses expire because of technical or bureaucratic issues. In the days that followed, we started texting each other and talking on the phone daily about this issue, until eventually a deal was made between Israel and South Korea, and Pfizer facilitated the transfer. Bennett would call me or text me regularly to discuss the progress.

Our research collaboration with Israel became the most reliable scientific tool the world had to assess the safety and efficacy of our vaccine and the impact of vaccination programs on the health and socioeconomic indexes of a country. Every Tuesday, the Israeli scientists shared with our experts weekly data analyzed in multiple different ways. Due to Israel's high level of vaccination rates, its outstanding electronic health records system, and its extensive program of testing, we were able to have a very accurate picture of the performance of our vaccine and observe any potential changes in a timely manner. The results were amazing, and Israel was able to completely open its economy and lift most restrictions while maintaining a very low number of cases, predominantly within the unvaccinated population. But Israelis never stopped testing for COVID-19, monitoring the disease among their citizens, and sharing their data with us, as we had agreed when we signed our research collaboration.

In late June, we started noticing a rapid rise of cases across the country. Because of our close collaboration and the excellent system we had put in place, we were able to quickly identify whether the cases were among unvaccinated people or were breakthrough infections (infections among vaccinated people), and within the breakthrough population, the personal characteristics of those who were being infected, including symptomatic and asymptomatic cases. This was critical. The rise in infections was occurring with the introduction of the more infectious delta variant, and what we needed to determine quickly was whether the upsurge in cases was caused by lack of efficacy of the vaccine against this variant or the gradual loss of efficacy over time. This had critical implications not just for Israel but also for the world, because if the former was true, then a new vaccine tailor-made to prevent the delta variant needed to be developed immediately. Our analysis revealed that the efficacy of the vaccine was waning over time and most specifically among the individuals vaccinated in large numbers beginning in December. However, among those who had received the vaccine only recently, it was holding up strongly at preventing both infections and severe disease due to delta.

At the beginning, the waning of efficacy was resulting in only a low level of infections and mild disease. Unfortunately, as time passed and the number of reinfected individuals grew, the rate of hospitalizations and severe disease cases also started to increase in the most vulnerable populations that were first vaccinated, making the question even more urgent of whether a booster dose was needed. Our studies evaluating the impact of a booster dose were ongoing, and the first results were very positive in both the safety and efficacy domains.

With data coming out of Israel on the waning of immunity, starting approximately six months after the second dose, I strongly believed that it was our moral imperative to share this information more widely. We had vowed to always be transparent, and we kept this promise throughout the pandemic. This helped us build a very high level of trust among the public. Now, for the first time, we learned information that raised concerns about our vaccine's duration of protection. Sticking to our principles to share information publicly, I asked my team to prepare a press release on July 8 detailing our findings from Israel.

"As seen in real world data released from the Israel Ministry of Health, vaccine efficacy in preventing both infection and symptomatic disease has declined six months post-vaccination, although efficacy in preventing serious illnesses remains high," we said in this joint Pfizer/BioNTech release. "Additionally, during this period the Delta variant is becoming the dominant variant in Israel as well as many other countries. These findings are consistent with an ongoing analysis from the Companies' Phase 3 study."

At the same time, we informed the public about our encouraging data from the booster study: "Initial data from the study demonstrate that a booster dose given 6 months after the second dose has a consistent tolerability profile while eliciting high neutralization titers against the wild type and the Beta variant, which are 5 to 10 times higher than after two primary doses."

Finally, we stated, "That is why we have said, and we continue to believe that it is likely, based on the totality of the data we have to date, that a third dose may be needed within 6 to 12 months after full vaccination. While protection against severe disease remained high across the full 6 months, a decline in efficacy against symptomatic disease over time and the continued emergence of variants are expected. Based on the totality of the data they have to date, Pfizer and BioNTech believe that a third dose may be beneficial to maintain the highest levels of protection."

In the past politicians and governments would pressure us to share data and learnings more quickly than we were comfortable. In this case, we believed that we were sharing timely data that could be critical for governments around the world to make significant public health decisions. To our surprise, our release set off a firestorm with US health officials.

In response to our statement, the CDC and the FDA took the unprecedented step before asking to see our data to put out their own statement that refuted our release's conclusion that boosters would be required. It is understandable that when complex public health challenges, politics, and science intersect, things can get messy. We have always tried to keep it simple—follow the science wherever it takes us and be candid with the public about what we are seeing. That's what we did when we made the booster news public.

I believe that the US officials got concerned that our announcement would increase vaccine hesitancy. We saw it differently. Our evidence to date indicated that the vaccine was still effective at preventing hospitalization and severe illness even after several months, but that protection against infection waned. Most important, we were announcing that we had a possible solution in hand to address waning infection efficacy, and the more modest decrease in effectiveness against severe illness.

In retrospect, we should have given the US agencies a heads-up before making our findings public. This was a mistake on our side. I called both Tony Fauci and Jeff Zients to apologize for catching the administration off guard. I emphasized, though, that our scientists strongly believed in the validity of our conclusions, and I offered to schedule a meeting where a group of Pfizer scientists could brief the appropriate US health officials on the totality of the data we had at our disposal. Both of them accepted happily to do that. This meeting happened two days later and was very productive. Tony Fauci, Francis Collins, Rochelle Walensky, Janet Woodcock, Peter Marks, and David Kessler were among the many who attended this meeting from the US government's side and discussed at length our data with Mikael Dolsten, Kathrin Jansen, Luis Jodar, and other Pfizer scientists. Similar meetings occurred with the European health authorities and the health authorities of other countries.

In the weeks that followed, the Israeli health officials became more concerned with the rise of more breakthrough cases and started analyzing more and more data. Bennett would text or call me frequently to ask for information on our booster studies and to inform me about their own findings. Similar to his predecessor, Netanyahu, he was very hands-on and extremely knowledgeable. He was very decisive. By the end of July, Israel's scientific committee of experts recommended booster vaccinations for people over sixty years old. The real-world evidence was helping them to inform a real-world response in record time. After studying the emerging data, Israel began a booster program with our current vaccine.

On August 3, Bennett shared with me a very important study that was conducted leveraging the computerized database of Maccabi Healthcare Services, the largest healthcare organization in Israel. The study assessed the correlation between time from vaccine and incidence of breakthrough infections and was confirming what we were saying. The risk of infections was multiple times higher for early vaccine recipients compared with those vaccinated later. By mid-August, the scientific committee recommended lowering the age of those entitled to receive a booster to fifty years old. As the first country in the world to begin its booster program to the general population in earnest, Israel was again able to provide invaluable data to the rest of the world on the effectiveness of boosters, including their capability to restore vaccine efficacy to the levels observed in our Phase 3 clinical program and the sustainability of their protective effects over time. In an analysis that was conducted between August 8 and August 21 in Israel, the vaccine efficacy of those who had received the third dose twelve days ago was restored to 93 percent for infections and 97 percent for severe illness. The relative risk of getting infected from COVID-19 due to delta was 11.5-fold lower for those receiving the booster compared to those that had two doses of the vaccine.

By August, the US administration announced that boosters were available for immunocompromised individuals vaccinated six to eight months before and, in a very high-profile appearance of Jeff Zients, together with Tony Fauci, Rochelle Walensky, and Janet Woodcock, announced their intention to roll out a broader booster program on September 20, 2021.

11

The Science of Trust

"The greater the difficulty, the more glory in surmounting it. Skillful pilots gain their reputation from storms and tempests."

—Epictetus, AD 50–135

THE STORY OF OUR RESEARCH, development, manufacturing, and distribution would be incomplete without the story of public perception. What if we had accomplished these breakthroughs only to discover that the public would refuse the shot for lack of trust in the industry, the company, or the science itself?

Trust is bedrock, foundational to a patient's decision to walk into a healthcare facility and accept an injection. Consumers hope the clothes they buy look good on them, that the car they purchase will drive smoothly for years to come. But hope can be insufficient when it comes to your body and your life. Merriam-Webster defines "trust" as an "assured reliance on the character, ability, strength, or truth of someone or something." Our messages, to trust science and to understand that science would win, became the centerpiece of our public positioning, and felt like a sacred duty to uphold. But threats to undermine trust were very real.

In July 2020, both the Pfizer and the Moderna vaccines were showing promise in early clinical trials. Both vaccines would be in Phase 3 trials during the fall. On July 31, 2020, Anthony Fauci appeared before the House Committee on Oversight and Reform's Select Subcommittee on the Coronavirus Crisis. He testified that he was "cautiously optimistic" the US would have a safe and effective vaccine by the end of the year. He cautioned, however, that the vaccine would become available to Americans in phases, not immediately. Around the same time, speculation swirled that President Trump, his reelection bid then in full swing, might fast-track a COVID-19 vaccine in hopes of reaching voters before election day. The *Financial Times* reported on August 30 that the president's FDA commissioner, Stephen Hahn, had told the newspaper in an interview that he was "willing to bypass the normal approval process." To make matters worse, computer hackers, propaganda, and misinformation campaigns fueled misunderstanding and confusion.

That summer, our executive team thought deeply about something bigger than politics: the public's trust. We were aware of attitudes toward the pharmaceutical industry, which competed with tobacco and government for the bottom rung of reputational surveys. The industry had done a lot of things in the past to lose trust from people. As a result, we also had to be concerned about our own company reputation and the reputation of the vaccine itself. Earning and maintaining the public's trust were essential. We know that a strong reputation is earned in drops but lost in buckets. We must earn trust every day because our patients expect it.

With COVID-19 infection rates on the rise and an imperative vaccine on its way, communications and government affairs would be essential and would play a major role in our Lightspeed deliberations.

We began immediately and aggressively to push back on notions that the vaccine would skip or short-circuit Phase 3 trials. We were moving fast, but not cutting corners. The heated political debate was not helping. For me, the straw that broke the camel's back was a tweet from President Donald Trump about the FDA. President Trump would ask me in every call I had with him during this period if we had problems with the FDA. It felt as if he were trying to get a statement from me that the FDA was doing something bad. In all cases, my answer was very clear: the FDA was doing an exceptionally good job in collaborating with us on this project. However, on Saturday, August 22, he tweeted:

"The deep state, or whoever, over at the FDA is making it difficult for drug companies to get people in order to test the vaccines and therapeutics. Obviously, they are hoping to delay the answer until after November 3rd. Must focus on speed, and saving lives!"

I became seriously concerned with this turn. If people started doubting the FDA's integrity, how would they trust a vaccine that would be approved by them? If it was approved before the elections, some might think that this was the result of the political pressure from the White House. If it was approved after the elections, others might think that this was the result of the political pressure from the Biden campaign. Either case would not help build people's trust in the process and the vaccine, and would bring more bad news for public health. We had to do something.

During the crisis, I became quite close with Alex Gorsky, the chairman and CEO of Johnson & Johnson. The two of us would talk on the phone regularly on weekends and discuss the situation with COVID-19 and the challenges we were faced with. I called Alex to discuss my concerns about the political environment and the political pressures. During our discussion we agreed that we must speak up. It would be a good idea to write a statement assuring the world that we would not cut any corners.

"Let me draft something and send it to you," I told Alex.

I called Sally Susman, and over one weekend we resolved to draft a public pledge that biopharma leaders, including myself, would sign to unite in solidarity for science and with regulators charged with the public safety. Politicians and pundits might advocate for weakening the scientific approach, but we would not. In the pledge, we wrote, "The agency requires that scientific evidence for regulatory approval must come from large, high quality clinical trials that are randomized and observer-blinded, with an expectation of appropriately designed studies with significant numbers of participants across diverse populations." We pledged to always make the safety and well-being of vaccinated individuals our top priority; to continue to adhere to high scientific and ethical standards regarding the conduct of clinical trials and the rigors of manufacturing processes; to only submit for approval or Emergency Use Authorization after demonstrating safety and efficacy through a Phase 3 clinical study that is designed and conducted to meet requirements of expert regulatory authorities such as the FDA; and to work to ensure a sufficient supply and range of vaccine options, including those suitable for global access.

I sent the pledge in draft form to Alex, who in a few hours returned the draft with a few improvements. The pledge was ready. We split up names of other biopharma CEOs to approach to take the pledge. I called half of them, and Alex called the other half. They all agreed. Nine CEOs signed on—those from AstraZeneca, BioNTech, GlaxoSmithKline, Johnson & Johnson, Merck, Moderna, Novavax, Pfizer, and Sanofi. On September 8, we issued the statement publicly, placing it as full-page ads in fourteen of the top daily newspapers in the United States. It sent a loud and clear message that the virus was the enemy, and that our industry and the FDA were aligned on public safety first and foremost. The pledge silenced the debate over undermining Phase 3 trials and kept the emphasis on safety and effectiveness. We were absolutely committed to seeing our vaccine candidate fully tested and analyzed before its authorized use.

We didn't stop there. Our decision to publish Pfizer's highly detailed clinical trial protocol also grew from this effort. It was hotly contested behind closed doors by many of our scientists.

"This is a highly scientific document that only scientists understand. What is the meaning of making it public? And are we creating an obligation to publish protocols for all our studies from now on?"

I knew that their concerns were rational, but we determined that transparency would help to build vaccine confidence. The *British Medical Journal* (*BMJ*) described the move as "a rare opportunity for public scrutiny of these key trials."

Rod MacKenzie, Pfizer's Chief Development Officer, responsible for overseeing clinical trials, was a strong and persistent voice for patients and transparency. Sally Susman, our Chief Corporate Affairs Officer, reminded us that, with the public feeling so vulnerable, we had to take extra steps to achieve our goal of leading the conversation.

With Sally and her team's support, I stepped up my public profile. Local, state, tribal, and federal governments were anxious. Rather than delegate calls from public officials to staff, I took them myself. In May, the Pew Research Center found that Americans were more positive than negative about COVID-19 news coverage. They felt media coverage of the crisis benefited the public. A later poll conducted by the Robert Wood Johnson Foundation and the Harvard School of Public Health found that Americans trust nurses, healthcare workers, and doctors more than public health institutions and agencies. There was a void. Rather than decline media interviews, I happily took them. This was a change. We had been practically in a bunker for years. In fact, the last Pfizer CEO to be profiled positively and prominently on a mainstream magazine cover was Bill Steere, Jr., who retired from the company in 2001. To build trust, I needed to become more visible and be available to answer hard questions from reporters and from elected leaders. We moved from simply trying to manage inquiries to actively participating in the daily news cycle. In the early days of the pandemic, I agreed to an extensive interview with *Forbes* about plans for a vaccine. In it, critics said my timeline was unrealistic. The cover story went in depth to show why we felt otherwise. The lockdown was so severe and sudden back then that in one photograph I'm in my garage, which was hastily turned into a makeshift studio; in another, the photographer stood outside my window, my arms were crossed with determination, and the trees outside my home reflected on the glass.

Over the summer, we launched a public service campaign, "Let's undo underrepresented diversity in clinical trials," which encouraged diverse communities to participate and be represented. On April 15, in the depths of the pandemic, we aired a sixty-second ad, *Science*

Will Win, on the *Today* show as a rallying cry for science and the scientists working on vaccines and therapies:

> At a time when things are most uncertain, we turned to the most certain thing there is, science.
>
> Science can overcome diseases, create cures and, yes, beat pandemics.
>
> It has before. It will again.
>
> Because when it's faced with a new opponent it doesn't back down, it revs up, asking questions till it finds what it's looking for.
>
> That's the power of science.
>
> So we are taking our science and unleashing it, our research, experts and resources, all in an effort to advance potential therapies and vaccines.
>
> Other companies and academic institutions are doing the same.
>
> The entire global scientific community is working together to beat this thing, and we're using science to help make it happen, because when science wins, we all win.

The closing image is one of our workers in the lab: "We thank all the scientists working relentlessly in both our labs and labs around the world to end this global health crisis." That fall, we took viewers behind the scenes in a four-minute episode, *No Stone Left Unturned in the Fight Against COVID-19*, which featured our scientists describing in their own words the work underway.

On Friday, October 16, just two weeks before the presidential election, I penned an open letter addressed to the billions of people and thousands of leaders around the world eagerly awaiting news of a vaccine. I wanted to make it clear that while we were nearing an important data readout on our Phase 3 trials, we would apply for Emergency Authorization Use only after safety milestones were achieved in the third week of November, after the elections. The letter became the most-read page ever on the Pfizer website.

On the first anniversary of the WHO declaring the pandemic, we broadcast a documentary, *Mission Possible*, with National Geographic to give the public a behind-the-scenes look at our work and a chance to see the faces and hear the voices of those who'd made the vaccine possible. In many ways, this book you are reading is part of that effort.

Being more open, more transparent, was the right decision, and it will continue to be our policy. As the vaccine became increasingly available in 2021, an Axios Harris Poll found that the pharmaceutical industry's reputation had doubled in positivity from prepandemic 2019, now rivaling tech and manufacturing. Our own research showed that our favorability surged. Our overall trust saw significant gains, breaking away from the pack in our industry. In time, everyone knew what a spike protein was. Pfizer entered pop culture, becoming a frequent subject of both praise and humor on *Saturday Night Live*.

AdAge wrote that "Pfizer has been, far and away, the big winner in vaccine brand popularity." The paper's five strategies for brands to succeed in a crowded market emphasized our visibility and transparency, and our inclusion of consumers in the product itself.

It's been meaningful and gratifying, to both me and our team, to see Pfizer's name appear on "best of" lists and to receive accolades. Popularity is fleeting, but a commitment to public trust is permanent. That commitment must be clear, consistent, and uncompromising. It must be shared by both public- and private-sector organizations. The media also plays a role. A January 2020 analysis of media coverage and the reputation of the pharmaceutical industry found that stories focused on business more than science, including mergers and acquisitions, restructuring, and financial reports. The pandemic has shown us that more and deeper coverage of science is urgently needed.

With great humility, I accepted the Appeal of Conscience Award on March 22, 2021. In my opening remarks, I acknowledged Rabbi Arthur Schneier, the founder, for creating an institution that models how human beings should treat one another, a living testament to the power of love, a powerful force for good in our world. It is through love that we foster mutual understanding that every person deserves to be seen, heard, and cared for. That is the fertile ground in which trust takes root and grows.

If the vaccine taught us anything about public perception, it is that our purpose—Breakthroughs that change patients' lives—is the only pathway to a sound reputation. And we will guard it with vigilance.

12

A Pro-Patient, Pro-Innovation Agenda

"What you leave behind is not what is engraved in stone monuments, but what is woven into the lives of others."

—*Pericles, 495–429 BC*

ON MAY 25, 1961, PRESIDENT Kennedy told Congress, "I believe that this nation should commit itself to achieving the goal, before this decade is out, of landing a man on the moon and returning him safely to the earth." Eight years later, on July 24, 1969, that mission was accomplished. Yet, in many ways, the moonshot continues as humankind explores the cosmos beyond the moon, and President Biden challenges the nation to embrace a science-based future.

Similarly, our moonshot also continues. On May 4, 2021—six months after our breakthrough—I told colleagues and stakeholders during our quarterly earnings report that I could not be prouder of Pfizer. On that day, we were on track to produce 2.5 billion doses of COVID-19 vaccine by end of the year, and 3 billion in 2022. We'd also proven that we were not a one-hit wonder. Excluding the revenue earned from our COVID-19 vaccine, Pfizer's revenues grew 8 percent operationally in Q1 2021. In fact, we also achieved several important clinical, regulatory, and commercial milestones. We had made a tremendous gambit to find a safe, effective vaccine. Science had won.

We had learned a lot about what it takes to be passionate and highly effective in driving innovation and impacting the lives of so many people. In the preceding chapters, I've explored the science, technology, and business approaches we pioneered under tremendous pressure. With COVID-19, everyone was at risk, so everyone was aligned in finding a cure. But most diseases do not affect everyone, which too often leads to a lack of alignment on how to approach healthcare, resulting in inefficiency and inequity. In this final chapter, I look ahead to how we might apply what we learned to future innovations and patient advances in a five-point recommendation.

1. Improve Access and Reimbursement Practices

We must drive policies that reduce out-of-pocket expenses for patients. Costs have grown, and as a society we are putting more costs on the patient. In the United States, Medicare Part D, for example, has no caps on out-of-pocket expenses, so a patient who may need bespoke treatments such as immuno- and gene therapies could be on the hook for more than 5 percent of expenses, no matter how high those expenses go. These expenses become consequential for many families. In some cases, biopharma invests in research focused on discovering innovative medicines that may serve only a small group of patients afflicted with a specific condition. Because of the small number of patients, treatments can be expensive. How can we as a society and an industry share in those costs to ensure the burden is not unjustly placed on the patient?

Also, how can we ensure that patient inputs are appropriately weighted and considered by payers? One approach is to increase participation by patient advocacy groups in what is known as Health Technology Assessment (HTA) decisions. HTAs are the mechanism by which governments and other healthcare payers evaluate the impact of a particular medicine or technology. We need to increase the number of HTA bodies that have a specific criterion or process to evaluate patient evidence. Europe's HTAs are a complex formula that runs costs and benefits through a wringer in order to spit out the perceived value of the medicine. Only that formula does not account for patient benefits, like whether a parent can continue to care for their child, or a colleague can continue to benefit from the dignity that comes from going to work. It only examines, under the most rigid formula, direct costs to the healthcare system. We believe the wellness and productivity of the patient should be taken into consideration.

We also need to advance the use of value-based pricing and payment models. To do so means reducing barriers to the adoption of value-based agreements. We can get there faster by working with governments to cocreate new pricing and payment models. We can demonstrate that *x* dollars in savings and *y* dollars in added productivity flow from a certain medicine or treatment. A patient's life is extended. A patient's stay in the hospital is reduced. How do we better account for these real-life outcomes?

While wealthy countries' health systems often discount the benefits that medicines bring to patients, these systems are often well funded but allocate resources to less efficient parts of the healthcare system. But in many low- and middle-income countries, the basic infrastructure to fund healthcare is inadequate, meaning many innovative medicines are out of reach for patients in these countries. It is critical that governments and the private sector collaborate to close this financing gap. That is why we have worked to promote public-private partnerships to develop commercial health insurance systems that can supplement the basic healthcare financing and reimbursement entities across the developing world. For example, we recently partnered with a Chinese insurance company, Ping An, and the city of Nantong in China to develop a pilot program to offer commercial health insurance to one million individuals in that city. The policies will supplement the basic medical insurance offered to Chinese citizens and be offered without regard to preexisting conditions, while providing access to innovative oncology and rare disease medicines to patients who would otherwise have no opportunity to access these breakthroughs.

In every country where we operate, we are working with patients and their advocacy organizations to create and enact creative policies to improve access and reimbursement.

2. Build Support for Intellectual Property

In preceding chapters, I've shared some of my frustrations over the challenges we've faced and are addressing about intellectual property protection. At the time of this writing, the risk to COVID-related patents is real and pending.

Going forward, we must build a basic understanding across the globe of the necessity of patents. I worry that we haven't found the right message yet to educate on this crucial point and often sound like a pamphlet from a pharmaceutical trade association when we speak of patents.

The basics are straightforward: Intellectual property (IP) is a key element in protecting a variety of different innovations. Within a legal framework, IP rights, particularly patent rights, allow an inventor to benefit from their creation by giving them control over how their intellectual property is used for a limited period of time. Governments establish and enforce intellectual property protection primarily to stimulate innovation in knowledge-intensive fields and encourage the production of useful goods and services arising from knowledge-based inventions. The establishment and protection of IP rights reflects a balance intended to advance the interests of inventors and society. It's that simple—take away the incentive, you take away the innovation.

I know from my background as a scientist and years as a business leader that strong IP systems foster an innovative culture, where innovators can develop new products and technologies knowing that their inventions and creativity are secure—and they can securely share such knowledge and inventions with partners and collaborators. It enables collaboration between biopharmaceutical innovators, governments, universities, and other research partners to speed up progress on the most pressing unmet medical needs. IP will continue to play a crucial role long after this pandemic is over, to ensure that the world is prepared with innovative solutions for future global health crises, in addition to other pressing healthcare needs.

3. Nurture the Future of Technology and Artificial Intelligence

The "tech-celeration" of our company and industry will continue to be a priority. Digital research and development is a rapidly evolving ecosystem with the potential to reduce barriers to innovation and expand the market for novel medical products and therapeutics. Artificial intelligence/machine learning (AI/ML) and connected devices and sensors represent the most commonly leveraged technologies for digital R&D. The area with the most promise is in the field of clinical trials—spanning the full spectrum of trial design, trial operation, and data collection—but those technologies are also being used in the discovery R&D processes and in manufacturing and life cycle management.

Digital transformation is allowing us not only to think big but also to dream big—to ask ourselves, *What if?* What if data and digital tools could reinvent prevention, patient empowerment, and the pace of both R&D and manufacturing? What if this confluence of science and technology were harnessed appropriately for better health outcomes?

For example, we can imagine matching patients with customized therapies, maximizing their benefit, and improving the patient's quality of life. This could enable long-term tracking of patients to ensure they are achieving the health outcomes that these therapies promise, and this rich data universe could lead to new discoveries and refined therapies. Combining the power of breakthrough medicines with digital therapeutics could improve health outcomes for patients and better care continuums for healthcare professionals.

Curative therapies could lower the overall impact and cost that chronic diseases have on healthcare systems. This could change the way we treat patients and impact how healthcare providers organize care and its delivery. Developing, marketing, and pricing these curative treatments will require the biopharma sector to adopt new data and analytic capabilities. New mRNA technology could drive prevention and early detection of disease.

The breakthroughs we have made with mRNA vaccine technology may allow for one vaccine to provide protection for multiple diseases, decreasing the number of shots needed for common vaccine-preventable diseases. In addition, cancer research is studying how to use mRNA to trigger the immune system to target specific cancer cells. We could leverage advanced data analytics and High Performance Computing power to identify patterns related to the causes and early markers of disease—and drive interventions to prevent disease altogether. This also supports a more equitable health system around the world as physicians are quickly provided access to relevant information through digital channels.

4. Empower Patients

What if patients were empowered—choosing when and how to engage and stay connected for wellness and better health outcomes? A platform that enables consumers/patients to easily gather, access, and manage their own health information would allow them to continually assess their own wellness and understand the most important drivers underlying their health. Personalized coaching and diagnostics would enable patients to maximize their own health, prevent disease, and ultimately access the right level of care when and where they need it to maximize positive outcomes when grappling with disease. Digital solutions could help patients who want to share information with their healthcare providers (leveraging biosensors for continuous, real-time physiological information via noninvasive measurements of biochemical markers) by enabling their doctors to contact them when the equivalent of a "check engine light" was triggered.

The ability for patients to leave a sample outside their own front door, track it, and receive a notification when the diagnostic result is ready could drive greater convenience, potentially speed diagnosis, and ensure earlier treatment for better health outcomes. We could drive greater diversity in, and accessibility to, clinical trials if we were able to seek out individual patients for trials, rather than the patient, often overwhelmed with a new diagnosis, having to seek out the trial. The ability for care/treatment journeys to be delivered more like travel itineraries could allow patients to better understand the time and financial commitments required at each leg of the journey. Allowing patients to easily share these itineraries with their loved ones (e.g., family, friends, local support services) as they navigate treatment could drive greater adherence to treatment and better outcomes. Patients could take better control of their health if they were able to enlist their personal devices to help manage their treatment journey. What if the patient's preferred device could coach them on not just when to leave for a medical visit but also what they can or cannot eat the night before, and what they might want to have on hand to provide comfort after the visit? What if a patient could signal that he or she was confused and receive help via machine or a specialized expert (e.g., pharmacist) at any time, such as reminders of what pill is which, what you can or cannot eat or drink, or what a test result means? What if the information a patient undergoing treatment shared influenced how members of the healthcare ecosystem were rewarded (ePRO for VBHC (value-based-healthcare) delivery)—rewarding therapies and delivery models that made meaningful differences in patients' lives, broadcasting where it did not make a difference and highlighting that unmet need as future R&D opportunities for any interested in listening? We could enable patients already providing their data for a trial to become citizen-scientists. Patients could be more involved in advancing the research they care most about (e.g., data donations), and they could signal their availability to provide more data or different data to enable the science. Providing rewards to consumers/patients aligned to the value their data helps create (e.g., shareholder in a new discovery) could drive engagement and lead to more scientific breakthroughs. The human impact of these possibilities is huge and truly excites me.

5. Never Stop Innovating

Rod MacKenzie, Pfizer's Chief Development Officer, builds on these "what if" questions when he asks: "Why just COVID-19?" Rod is a passionate scientist from the west of Scotland with the brogue to prove it. He went to college locally at the University of Glasgow and studied organic chemistry at Imperial College in London and Columbia University in New York. Like everyone on my executive leadership team, Rod had specific goals during the pandemic. I assigned him the goal of leading a regulatory revolution. As the vaccine inoculated millions and then billions, his mantra became "Why just COVID-19?" Why can't we apply our learning from the pandemic to cancer and other diseases? Why go back to expectations that it will take two months just to get a meeting with regulators? Why can't we parallel process more research, development, and manufacturing processes? People are dying every day from a host of diseases that must someday soon become preventable and treatable.

It is time to take immediate actions to improve clinical trials and better serve all patients. Extraordinary times deserve extraordinary actions. But if we look beyond the current pandemic, we are confronted with some uncomfortable questions: How can we take such exceptional action for COVID-19 patients, but not for patients with cancer, life-limiting autoimmune conditions, fatal genetic disorders,

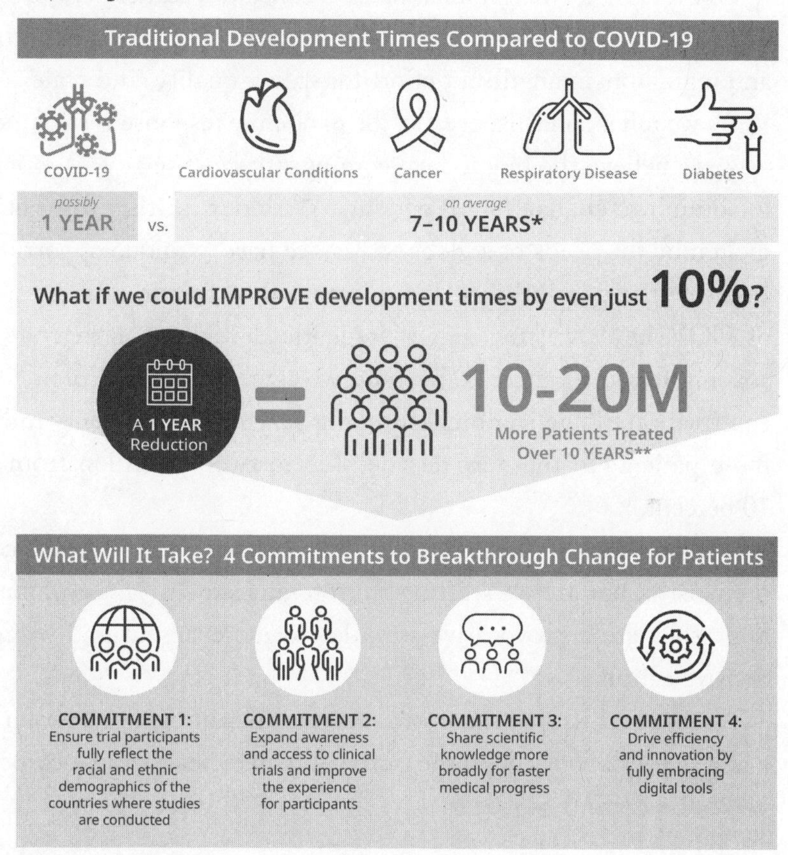

Applying the experiences and clinical development process improvements from the COVID-19 vaccine program, Pfizer made four public commitments in September 2020 to ensure that all patients can benefit from scientific breakthroughs as quickly as possible
Image courtesy of Pfizer Inc.

or a myriad of other major medical needs? At the time of this writing, cumulative global mortality from COVID-19 has surpassed five million, a horrifying toll, and it will continue to rise. But each and every year, about seventeen million people die from cardiovascular disease and ten million from cancer. Are these patients somehow less deserving?

Of course not. And yet it has taken the medical, economic, and social impact of a global pandemic to align sponsors, clinical researchers, and regulators in an all-out effort for speed, quality, and scale.

As we reflect on this, even as the pandemic response is still underway, we believe the biomedical community must seize this moment to commit to change for all patients. With our existing conventions molten by the needs of a global crisis, we have a unique opportunity. Imagine the possibilities. If we keep only a fraction of the speed of COVID-19 vaccine and therapeutic development that we are striving for, just a mere 10 percent, we estimate that the biomedical enterprise could deliver breakthroughs serving ten to twenty million more patients in the next decade. Ten to twenty million from just 10 percent.

The proposed actions discussed here are not intended to be comprehensive, nor are we starting from scratch in any of them, but we hope they are a provocative stimulus to more urgent and comprehensive action.

I'm hopeful that in the postpandemic environment we can create a new conversation around the future of human health and incorporate and build upon these ideas.

Epilogue

ON AUGUST 23, 2021, I was opening our weekly Monday-morning executive leadership team meeting. The team and I were gathering and settling in when an email crossed at 8:59 a.m. from Elisa Harkins Tull, Senior Director, Global Regulatory Affairs—Vaccines in our Collegeville, Pennsylvania, facility, with the header "COMIRNATY (COVID-19 Vaccine, mRNA) BLA APPROVED."

Elisa said it well:

Dear All,

I am writing, on behalf of the Pfizer and BioNTech Regulatory Team, to inform you that the Biologics License Application (BLA) for COMIRNATY® (COVID-19 Vaccine, mRNA) has been APPROVED for individuals 16 years of age and above!!! This represents the first approval for full licensure of our COVID-19 vaccine. The Approval Letter is attached.

The speed at which this was accomplished has been astounding. Not because there was any less of a burden of evidence than would be required for approval of a BLA for a vaccine and not because any corners were cut for the review. The speed was accomplished due to the dedication, determination, and extreme number of hours worked by an extensive amount of Pfizer, BioNTech and US FDA colleagues.

- The initial roll of the BLA was submitted on May 6
- The final roll of the BLA was submitted on May 18, which started the clock

- The BLA was granted Priority Review Designation with an Action Date of January 18, 2022
- The BLA was approved TODAY!

In addition, Emergency Use Authorization was granted for COMIRNATY for use in individuals 12 through 15 years and to provide a third dose to individuals 12 years of age and older who have been determined to have certain kinds of immunocompromise. The reason for this is to allow for these populations to be able to be vaccinated with product labeled as COMIRNATY once that product enters the market. The Letter of Authorization is attached.

For all of the colleagues who have contributed to this, your work has been absolutely monumental!

CONGRATULATIONS to ALL!!!

Elisa's note went to 431 people in the company. Earlier that year, as part of our simplicity initiative and to unclog our inboxes, I set a new norm that we do not use "reply all." On this incredible occasion, many people ignored my dictate and responded to everyone with messages of pride and joy. I couldn't complain.

Of course, this full approval was anticipated, and by no means a surprise. Still, I didn't anticipate how historic it would be and how deeply moving it would feel. Immediately, I issued a letter of thanks to all Pfizer colleagues, then I sent another to the Lightspeed team members, and some personal ones to individuals who had gone above and beyond. I also sent a note to my executive leadership team, who had been extraordinary.

Dear ELT,

What an incredible journey! Back in March of 2020, when we decided to pursue a vaccine for COVID-19, very few believed we would be able to deliver one in less than a year and secure full FDA approval in less than 18 months . . . but *you* did. You believed in our science. You believed in our people. You believed in our purpose. And you believed in one another.

Thank you for your leadership, your excellence and your friendship. We never could have achieved so much so quickly without your expertise, courage and sacrifice. The journey is far from over, but with this team leading the way, I am certain that science will win!

Thank you! Albert

Throughout the day, I received many phone calls, emails, and letters of thanks, as did my colleagues. Each of these communications touched me, whether it was the father who wrote to say he finally felt safe for his kids to be vaccinated or the entrepreneur who expressed awe that a large, traditional pharmaceutical company could have been on the cutting edge of science and moved with such urgency. That afternoon, I did several interviews, including one with NBC News anchor Lester Holt, with whom I had spoken many times over the previous months.

Finally, that evening I went home and had a moment of quiet. As I reflected, I knew that two truths would remain with me forever.

The first truth is that our breakthrough vaccine was the result of the rare combination of brilliant, cutting-edge science powered by the private sector alongside collaborative engagements with governments. It's fascinating to see that in the last year "mRNA" became a household word. Laypeople followed the science as if it were a popular sport. It's imperative that our society continues to respect and honor science. To me, the women and men who studied science and human health for years (and often decades) and have devoted their lives at the bench or scouring data in search of discoveries, often in obscurity, are heroes. We must continue to treat them as such. I look forward to the day when schoolchildren are as familiar with the names and faces of our leading scientists as they are with celebrities and sports stars.

I'm a true believer in scientists and am honored to have five outstanding scientific leaders on the Pfizer board, three of whom I have recruited to Pfizer since becoming CEO. We are better for their presence around our board table. I revel in the good news that medical school applications are on the rise. According to the American Association of Medical Colleges, "nearly two dozen medical schools have seen applications jump by at least 25%." Just as in the aftermath of 9/11 many young Americans chose to serve their country, today's graduates are selecting careers in science and medicine in response to the pandemic. I applaud them in their noble pursuit and cannot wait to see what they achieve.

"Science Will Win" became our mantra across Pfizer over these past eighteen months. We emblazoned the phrase on T-shirts and across our face masks, plastered it on the outside of our headquarters, and sometimes shouted it at the end of our Lightspeed meetings. I'm going to insist it remain our rallying cry.

Let us always remember that businesses have the power to make a positive difference. Corporations play a vital role as engines of creativity and avenues of opportunity. Commercial pressures require businesspeople to quickly discover and adapt new technologies and boost efficiency and productivity. Entrepreneurship and innovation go hand in hand.

I sometimes hear young people speak of the commercial world as "the dark side." I was young once, so I understand where this popular misconception comes from. My three decades in business have shown me the incredible strength and force for good that a well-resourced, soundly led, solutions-based company can be. That's why I left academia for Pfizer, and I've never looked back. Being able to pull the powerful levers at Pfizer to move quickly and independently was a critical factor in our success.

I'll continue to be an active member in organizations like the Business Roundtable and in other forums where I have the opportunity to meet with my fellow business leaders. None of us has a permanent lease on the corner office we inhabit. We must always stretch our thinking on how we can contribute to the greater good. We must demand major commitments for our companies and sacrifices of ourselves especially if we want to attract top talent.

And we must seek pathways for collaboration with governments around the world. We cannot allow extreme partisanship and rancor to drive us to bury our heads in the sand. I've considered it a great privilege when I had the opportunities to meet the elected leaders, from all political parties and governments around the world. I was never annoyed when my phone rang at an odd hour, with a request from an elected leader or health official trying to find vaccines for their people. As I've written in earlier pages, several of these individuals are now friends. As I'm a bit of a political and history enthusiast, I admit, I loved it.

Getting all these stars to align—the science, the private sector, and the governments—was the formula for our COVID-19 success and an imperative to resolve other crises in the future.

The second truth is that the epic approval achieved on August 23, 2021, was thanks to all the changes we made to strengthen Pfizer's culture in the years leading up to the vaccine, before we knew about the pandemic. It was not only the focus we placed on innovation or the investments we made in increased resources for digital technologies and research. Yes, those investments were important. But the most impactful change was the cultivation of a purpose-driven culture.

In all businesses, companies that stay true to their purpose perform far better than those that do not. When you align around purpose, you are sending very clear guidance to an organization with multiple layers. That clarity brings everyone much closer to what needs to be done and improves productivity. Now, take a situation like ours, where our purpose is to save lives—it's a noble one. That adds passion that people rally around. This culture that we created gave us the appropriate mindset and allowed Pfizer to move with the agility and speed of a small biotech and bring to the world a breakthrough vaccine that is dramatically changing the lives of so many. Whether you are a leader or a team member, I would encourage you to ask yourself the following three questions during every step of your journey:

- Am I being true to my purpose?
- Have I aimed high enough?
- Do I have the right mindset?

Every member of our team answered these questions affirmatively during our quest to make the impossible possible. Now everyone in the company also wants to pursue a moonshot for their area of human health. We dream of a future when a host of diseases will someday soon become preventable or treatable. All of us share an impatience, especially now that we know what's possible.

I could go on . . . and like so many of us at Pfizer, I will tell these stories for the rest of my life. What an honor it has been to be able to work so purposefully during this difficult time and through so much suffering.

I'd like to end where we began—with the eloquent foreword written for this book by former president Jimmy Carter, now ninety-seven. He's accomplished so much, including having authored over thirty books (which, having written only this one, I cannot even imagine). President Carter served his country in the navy and then became a successful businessman. He's a politician who achieved the highest office of the US, and is a true humanitarian and philanthropist, having founded the Carter Center, which Pfizer partners with to provide healthcare to the most vulnerable. He's volunteered for physical labor with Habitat for Humanity and for decades taught Sunday school. I admire him and take inspiration from his words: "I have one life and one chance to make it count for something . . . My faith demands that I do whatever I can, wherever I am, whenever I can, for as long as I can with whatever I have to try to make a difference." I couldn't agree more.

The events and experiences in the book have been faithfully rendered as I have remembered them, to the best of my ability. I've tried to retell my conversations in a way that evokes the real feeling and meaning of what was said at the time, but they are based on my recollection and are not intended to be a word-for-word documentation. Others have read the manuscript and, to the extent they have knowledge, have confirmed its accuracy.

At the time I wrote this in the summer of 2021, the world was still fighting COVID-19 and its variants. Pfizer leaders remained at the forefront of the effort to produce vaccines and support their use around the globe. My story captures only a moment in time of an ongoing and rapidly evolving pandemic.

Acknowledgments

I NEVER IMAGINED I'D WRITE a book. Leading Pfizer's effort to bring forward our COVID-19 vaccine created many unexpected opportunities—and responsibilities. One of those was to record our journey for future leaders and to document the triumph achieved by the nearly ninety thousand colleagues whose stories of perseverance, creativity, and sacrifice are told in these pages. It is for their legacy that I've authored *Moonshot.*

First and foremost, I offer my heartfelt congratulations to the Project Lightspeed members. In this bold endeavor to make the impossible possible, you managed to both aim high *and* hit.

To our COVID-19 vaccine partners at BioNTech, in particular Ugur Şahin and Özlem Türeci. I thank you, and consider it an honor to have traveled this road together.

To the Pfizer executive leadership team, who contributed to the vaccine while also delivering every day around the world on our purpose: *Breakthroughs that change patients' lives.* I can imagine no better team and am grateful for your partnership.

To the Pfizer board of directors, who provided wisdom and support throughout the pandemic and consistently over time. Thank you for being so resolutely behind our vision.

To the Pfizer team who contributed their time to this book: Eric Aaronson, Frank Briamonte, Andrea Christensen, Dana Gandsman, Ed Harnaga, Doug Lankler, Debra Mangone, Anneka Norgren,

Sally Susman, and Tiffany Trunko. Special thanks to Sally, my partner and prodder throughout this project.

To the dedicated team of professionals who move mountains every day in support of the office of the CEO, including in the support of the development of this book: Michele Bander, Dana Dotti, Steve Fascianella, Lily Hakim, Sonia Heidel, Yolanda Lyle, and Dorothy O'Mara.

My sincere thanks to all those who were interviewed to help account for posterity: Yasmeen Agosti, Tanya Alcorn, Payal Becher Sahni, Kim Bencker, Mois Bourla, Selise Bourla, Donna Boyce, Frank D'Amelio, Lindsay Dietschi, Mikael Dolsten, Phil Dormitzer, Lidia Fonseca, Bill Gruber, Ed Harnaga, Susan Hockfield, Angela Hwang, Kathrin Jansen, Luis Jodar, John Ludwig, Rod MacKenzie, Mike McDermott, Kevin Nepveux, Dara Richardson-Heron, Caroline Roan, Martina Ryall, Susan Schuman, Jon Selib, and John Young.

To my external advisors, who gave generously of their expertise: Greg Shaw at Clyde Hill Publishing, Hollis Heimbouch and the team at HarperCollins, Mollie Glick and the team at CAA, and Alan Fleischmann and his colleagues at Laurel Strategies.

To all my readers, please know that this work represents my memory of events during an incredibly pressured period. I've endeavored to make this work as accurate as possible. Any mistake in that regard is my own.

Appendix

Albert Bourla's "We Stand With Science" Open Letter to Colleagues

Subject Line: We Stand With Science
September 8, 2020

Dear Colleagues,

Over centuries, vaccines have saved millions of lives and altered the course of history. In fact, behind clean water, immunizations are the single most important health investment humans have made.

Pfizer has a rich history in vaccine research and development. For more than 130 years, we have played a pivotal role in eliminating or nearly eliminating deadly infectious diseases like smallpox and polio. We have designed novel vaccines based on new delivery systems and technologies that have helped in preventing bacterial infections. And today, we are hopeful that we, along with other biopharmaceutical companies, are on the cusp of making history again with the successful development of a safe and effective vaccine against COVID-19.

Our success in pioneering vaccines is rooted in our recognition that good science demands rigor, our commitment to patient safety, and our close partnership with regulators that are equally committed to scientific integrity. Together, these principles serve as our North Star and guide our work toward a potential mRNA vaccine and therapeutics against COVID-19. We will cut no corners in this pursuit.

In this spirit, Pfizer has signed a public pledge—along with a number of other vaccine researchers and developers—to protect the time-tested scientific processes and regulatory protocols that have ensured the safe delivery of medicines and vaccines to address patients' unmet needs. This show of solidarity complements the Five Point Plan that Pfizer issued in mid-March calling for unprecedented industry collaboration in fighting COVID-19, and I couldn't be prouder of our industry. I also couldn't be prouder of our company, as it was Pfizer that conceived of the pledge and invited our fellow industry leaders to stand with us.

Today, as always, Pfizer will **Stand With Science**, reinforcing our commitment to developing and testing potential vaccines for COVID-19 according to scientific principles, and not politics. And we call on everyone—including all of you—to join us in this commitment.

Sincerely,

Albert

Bibliography

Preface

Centers for Disease Control and Prevention. "Polio Vaccination." Vaccines and Preventable Diseases. Updated May 4, 2018. https://www.cdc.gov/vaccines/vpd/polio/index.html.

Dun & Bradstreet. "Array Biopharma Inc." D&B Business Directory. Updated January 1, 2019. https://www.dnb.com/business-directory/company-profiles.array_biopharma_inc.4784383beada038eadb66c2232df9ddd.html.

Hopkins, Jared S. "Mylan Deal Furthers Pfizer CEO's Bet on Patent-Protected Drugs." *The Wall Street Journal*. Updated July 29, 2019. https://www.wsj.com/articles/pfizer-to-merge-off-patent-drug-business-with-mylan-11564398516.

Kennedy, John F. "John F. Kennedy Moon Speech-Rice Stadium." Johnson Space Center. September 12, 1962. https://er.jsc.nasa.gov/seh/ricetalk.htm.

Kuchar, E., M. Karlikowska-Skwarnik, S. Han, and A. Nitsch-Osuch. "Pertussis: History of the Disease and Current Prevention Failure." In *Pulmonary Dysfunction and Disease*, edited by Mieczyslaw Pokorski, 77–82. Advances in Experimental Medicine and Biology. Cham, Switzerland: Springer International Publishing, 2016.

Mazzucato, Mariana. *Mission Economy: A Moonshot Guide to Changing Capitalism*. New York: Harper Business, 2021.

Merriam-Webster. "Moon shot." https://www.merriam-webster.com/dictionary/moon%20shot.

Murray, Jessica. "Covid vaccine: UK woman becomes first in world to receive Pfizer jab." *The Guardian*. December 8, 2020. https://www.theguardian.com/world/2020/dec/08/coventry-woman-90-first-patient-to-receive-covid-vaccine-in-nhs-campaign.

Pfizer. "Pfizer Reports Second-Quarter 2020 Results." Press release. Business Wire. July 28, 2020. https://www.businesswire.com/news/home/20200728005358/en/Pfizer-Reports-Second-Quarter-2020-Results.

World Health Organization. *Novel Coronavirus (2019-nCoV) Situation Report-1*. January 21, 2020. https://apps.who.int/iris/bitstream/handle/10665/330760/nCoVsitrep21Jan2020-eng.pdf?sequence=3&isAllowed=y.

World Health Organization. *Origin of SARS-CoV-2*. March 26, 2020. https://apps.who.int/iris/bitstream/handle/10665/332197/WHO-2019-nCoV-FAQ-Virus_origin-2020.1-eng.pdf.

Chapter 1

Acevedo, Nicole, and Minyvonne Burke. "Washington state man becomes first U.S. death from coronavirus." NBC News. February 29, 2020. https://www.nbcnews.com/news/us-news/1st-coronavirus-death-u-s-officials-say-n1145931.

Centers for Disease Control and Prevention. "In the Absence of SARS-CoV Transmission Worldwide: Guidance for Surveillance, Clinical and Laboratory Evaluation, and Reporting." SARS Home. Updated May 3, 2005. https://www.cdc.gov/sars/surveillance/absence.html.

C-SPAN. "President Trump Meeting with Pharmaceutical Executives on Coronavirus." March 2, 2020. https://www.c-span.org/video/?469926-1/president-trump-meeting-pharmaceutical-executives-coronavirus.

Greenwood, Brian. "The contribution of vaccination to global health: past, present and future." *Philosophical Transactions of the Royal Society B* 369, no. 1645 (June 2014). https://doi.org/10.1098/rstb.2013.0433.

Herper, Matthew. "In the race for a Covid-19 vaccine, Pfizer turns to a scientist with a history of defying skeptics—and getting results." *Stat.* August 24, 2020. https://www.statnews.com/2020/08/24/pfizer-edge-in-the-race-for-a-covid-19-vaccine-could-be-a-scientist-with-two-best-sellers-to-her-credit/.

Kathimerini. "Delphi Economic Forum postponed because of coronavirus fears." February 29, 2020. https://www.ekathimerini.com/news/250087/delphi-economic-forum-postponed-because-of-coronavirus-fears/.

LeDuc, James W., and M. Anita Barry. "SARS, the First Pandemic of the 21st Century." *Emerging Infectious Diseases* 10, no. 26 (November 2004). https://doi.org/10.3201/eid1011.040797_02.

Li, Shan, and Joyu Wang. "Wuhan Coronavirus Hospitals Turn Away All but Most Severe Cases." *The Wall Street Journal*. Updated February 5, 2020. https://www.wsj.com/articles/united-american-airlines-suspend-hong-kong-service-as-coronavirus-saps-demand-11580897463.

Office of Disease Prevention and Health Promotion. "Immunization and Infectious Diseases." Healthy People 2020. Updated June 23, 2021. https://www.healthypeople.gov/node/3527/data-details.%C2%A0Accessed.

Okwo-Bele, Jean-Marie. "Together we can close the immunization gap." World Health Organization Media Centre. April 22, 2015. https://apps.who.int/mediacentre/commentaries/vaccine-preventable-diseases/en/index.html.

Pfizer. "Kathrin U. Jansen, Ph.D." https://www.pfizer.com/people/medical-experts/vaccinations/kathrin_jansen-phd.

Pfizer. "Mikael Dolsten, M.D./Ph.D." https://www.pfizer.com/people/leadership/executives/mikael_dolsten-md-phd.

UNICEF. "UNICEF reaches almost half of the world's children with life-saving vaccines." Press release. April 26, 2017. https://www.unicef.org/press-releases/unicef-reaches-almost-half-worlds-children-life-saving-vaccines.

Vaccines Europe. *Improving Access and Convenience to Vaccination.* June 2018. https://www.vaccineseurope.eu/wp-content/uploads/2018/06/VE-Flu-Vaccination-Access-Pharmacies-0506018-FIN-FIN.pdf.

World Health Organization. "Middle East respiratory syndrome coronavirus (MERS-CoV)." Newsroom. March 11, 2019. https://www.who.int/news-room/fact-sheets/detail/middle-east-respiratory-syndrome-coronavirus-(mers-cov).

World Health Organization. "Vaccines and immunization." https://www.who.int/health-topics/vaccines-and-immunization.

Chapter 2

BioNTech. "BioNTech Signs Collaboration Agreement with Pfizer to Develop mRNA-based Vaccines for Prevention of Influenza." News release. August 16, 2018. https://investors.biontech.de/news-releases/news-release-details/biontech-signs-collaboration-agreement-pfizer-develop-mrna-based.

BioNTech. "Our Vision." https://biontech.de/our-dna/vision.

Cohen, Jon. "Chinese researchers reveal draft genome of virus implicated in Wuhan pneumonia outbreak." *Science.* Updated January 11, 2020. https://www.sciencemag.org/news/2020/01/chinese-researchers-reveal-draft-genome-virus-implicated-wuhan-pneumonia-outbreak.

Dormitzer, Philip. "Rapid production of synthetic influenza vaccines." *Current Topics in Microbiology and Immunology* 386 (2015): 237–73. https://doi.org/10.1007/82_2014_399.

Geall, Andrew, Ayush Verma, Gillis R. Otten, Christine A. Shaw, Armin Hekele, Kaustuv Banerjee, Yen Cu, et al. "Nonviral Delivery of Self-Amplifying RNA Vaccines." *Proceedings of the National Academy of Sciences* 109, no. 36 (September 4, 2012): 14604–9. https://doi.org/10.1073/pnas.1209367109.

Marshall, Heather D., and Vito Iacoviello. "mRNA Vaccines for COVID-19—How Do They Work?" *EBSCO Health Notes.* Updated March 16, 2021. https://www.ebsco.com/blogs/health-notes/mrna-vaccines-covid-19-how-do-they-work.

Pardi, Norbert, Michael J. Hogan, Frederick W. Porter, and Drew Weissman. "mRNA Vaccines—A New Era in Vaccinology." *Nature Reviews Drug Discovery* 17 (2018): 261–79. https://doi.org/10.1038/nrd.2017.243.

Pfizer. "Pfizer and BioNTech Announce Further Details on Collaboration to Accelerate Global COVID-19 Vaccine Development." Press release. April 9, 2020. https://www.pfizer.com/news/press-release/press-release-detail/pfizer-and-biontech-announce-further-details-collaboration.

Pfizer. "Pfizer and BioNTech Reach Agreement with Covax for Advance Purchase of Vaccine to Help Combat COVID-19." Press release. January 22, 2021. https://www.pfizer.com/news/press-release/press-release-detail/pfizer-and-biontech-reach-agreement-covax-advance-purchase.

Pfizer. "William C. Gruber, M.D., FAAP, FIDSA." https://www.pfizer.com/people/medical-experts/vaccinations/william_gruber-md-faap-fidsa-0.

Chapter 3

Centers for Disease Control and Prevention. "1918 Pandemic Influenza: Three Waves." Updated May 11, 2018. https://www.cdc.gov/flu/pandemic-resources/1918-commemoration/three-waves.htm.

The College of Physicians of Philadelphia. "Vaccine Development, Testing, and Regulation." The History of Vaccines. Updated January 17, 2018. https://ftp.historyofvaccines.org/content/articles/vaccine-development-testing-and-regulation.

David, Sharoon, and Paras B. Khandhar. "Double-Blind Study." In *StatPearls*. Treasure Island (FL): StatPearls Publishing, 2021. https://www.ncbi.nlm.nih.gov/books/NBK546641/.

US Department of Health and Human Services, Food and Drug Administration, Center for Biologics Evaluation and Research. *Development and Licensure of Vaccines to Prevent COVID-19: Guidance for Industry*. June 2020. https://www.fda.gov/media/139638/download.

US Department of Health and Human Services, Food and Drug Administration, Center for Biologics Evaluation and Research. *Emergency Use Authorization for Vaccines to Prevent COVID-19: Guidance for Industry*. May 25, 2021. https://www.fda.gov/media/142749/download.

US Food and Drug Administration. "Coronavirus (COVID-19) Update: FDA Takes Action to Help Facilitate Timely Development of Safe, Effective COVID-19 Vaccines." Press release. June 30, 2020. https://www.fda.gov/news-events/press-announcements/coronavirus-covid-19-update-fda-takes-action-help-facilitate-timely-development-safe-effective-covid.

Chapter 4

Hopkins, Jared S. "How Pfizer Delivered a Covid Vaccine in Record Time: Crazy Deadlines, a Pushy CEO." *The Wall Street Journal*. December 11, 2020. https://www.wsj.com/articles/how-pfizer-delivered-a-covid-vaccine-in-record-time-crazy-deadlines-a-pushy-ceo-11607740483.

Johnson, Carolyn Y. "Large U.S. covid-19 vaccine trials are halfway enrolled, but lag on participant diversity." *The Washington Post*. August 27, 2020. https://www.washingtonpost.com/health/2020/08/27/large-us-covid-19-vaccine-trials-are-halfway-enrolled-lag-participant-diversity/.

Pfizer. "About our Landmark Trial." 2021. https://www.pfizer.com/science/coronavirus/vaccine/about-our-landmark-trial.

Pfizer. "BioNTech and Pfizer announce regulatory approval from German authority Paul-Ehrlich-Institut to commence first clinical trial of COVID-19 vaccine candidates." Press release. April 22, 2020. https://www.pfizer.com/news/press-release/press-release-detail/biontech_and_pfizer_announce_regulatory_approval_from_german_authority_paul_ehrlich_institut_to_commence_first_clinical_trial_of_covid_19_vaccine_candidates.

Pfizer. "Commitment to Diversity." Coronavirus COVID-19 Vaccine Updates. 2021. https://www.pfizer.com/science/coronavirus/vaccine/rapid-progress.

Pfizer. "Our Values and Culture." 2019. https://www.pfizer.com/sites/default/files/investors/financial_reports/annual_reports/2019/our-purpose/our-values-and-culture/index.html.

Pfizer. 2016 Annual Review. February 2017. https://www.pfizer.com/sites/default/files/investors/financial_reports/annual_reports/2016/assets/pdfs/pfi2016ar-full-report.pdf.

Redd, Nola Taylor. "How Fast Does Light Travel? | The Speed of Light." Space.com. March 6, 2018. https://www.space.com/15830-light-speed.html.

Rottas, Melinda, Peter Thadeio, Rachel Simons, Raven Houck, David Gruben, David Keller, David Scholfield, et al. "Demographic diversity of participants in Pfizer sponsored clinical trials in the United States." *Contemporary Clinical Trials* 106 (July 2021): 106421. https://doi.org/10.1016/j.cct.2021.106421.

Treisman, Rachel. "Outpacing The U.S., Hard-Hit Navajo Nation Has Vaccinated More Than Half Of Adults." NPR. April 26, 2021. https://www.npr.org/sections/coronavirus-live-updates/2021/04/26/990884991/outpacing-the-u-s-hard-hit-navajo-nation-has-vaccinated-more-than-half-of-adults.

US Census Bureau. "QuickFacts." 2019. https://www.census.gov/quickfacts/fact/table/US/RHI725219.

Chapter 5

Bosman, Julie, Audra D. S. Burch, and Sarah Mervosh. "One Day in America: More Than 121,000 Coronavirus Cases." *The New York Times*. November 5, 2020. https://www.nytimes.com/2020/11/05/us/covid-one-day-in-america.html.

Dreisbach, Tom. "Pfizer CEO Sold Millions In Stock After Coronavirus Vaccine News, Raising Questions." NPR. November 11, 2020. https://www.npr.org/2020/11/11/933957580/pfizer-ceo-sold-millions-in-stock-after-coronavirus-vaccine-news-raising-questio.

Mueller, Benjamin. "U.K. Approves Pfizer Coronavirus Vaccine, a First in the West." *The New York Times*. December 2, 2020. https://www.nytimes.com/2020/12/02/world/europe/pfizer-coronavirus-vaccine-approved-uk.html.

Chapter 6

Pfizer. "Pfizer and BioNTech Announce Vaccine Candidate Against COVID-19 Achieved Success in First Interim Analysis from Phase 3 Study." Press release. November 9, 2020. https://www.pfizer.com/news/press-release/press-release-detail/pfizer-and-biontech-announce-vaccine-candidate-against.

Pfizer. "Pfizer and BioNTech Receive Authorization in the European Union for COVID-19 Vaccine." Press release. December 21, 2020. https://www.pfizer.com/news/press-release/press-release-detail/pfizer-and-biontech-receive-authorization-european-union.

Rosen, Bruce, Ruth Waitzberg, and Avi Israeli. "Israel's rapid rollout of vaccinations for COVID-19." *Israel Journal of Health Policy Research* 10, no. 6 (January 6, 2021). https://doi.org/10.1186/s13584-021-00440-6.

Sweet, Jesse, dir. "Mission Possible: The Race for a Vaccine." National Geographic CreativeWorks: Washington, DC, 2021. Aired March 11, 2021, on National Geographic. https://www.youtube.com/watch?v=jbZUZ9JYNBE.

US Food and Drug Administration. "Pfizer-BioNTech COVID-19 Vaccine." https://www.fda.gov/emergency-preparedness-and-response/coronavirus-disease-2019-covid-19/pfizer-biontech-covid-19-vaccine.

World Health Organization. COVID-19 Weekly Epidemiological Update. November 17, 2020. https://www.who.int/docs/default-source/coronaviruse/situation-reports/weekly-epi-update-14.pdf.

World Health Organization. "WHO issues its first emergency use validation for a COVID-19 vaccine and emphasizes need for equitable global access." Press release. December 31, 2020. https://www.who.int/news/item/31-12-2020-who-issues-its-first-emergency-use-validation-for-a-covid-19-vaccine-and-emphasizes-need-for-equitable-global-access.

Chapter 7

BBC News. "Covid-19: First man to get jab William Shakespeare dies of unrelated illness." May 25, 2021. https://www.bbc.com/news/uk-england-coventry-warwickshire-57234741.

Boucher, Dave, Kristen Jordan Shamus, and Todd Spangler. "Biden postpones visit to Pfizer facility in Portage until Friday." *Detroit Free Press*. February 17, 2021. https://www.freep.com/story/news/politics/2021/02/17/joe-biden-pfizer-facility-portage/6785063002/.

Burger, Ludwig. "BioNTech-Pfizer raise 2021 vaccine output goal to 2.5 billion doses." Reuters. March 30, 2021. https://www.reuters.com/article/us-health-coronavirus-biontech-target/biontech-pfizer-raise-2021-vaccine-output-goal-to-2–5-billion-doses-idUSKBN2BM1BW.

Bush, Evan, and Sandi Doughton. "Front-line medical workers get first doses of coronavirus vaccine in Seattle." *The Seattle Times*. Updated December 16, 2020. https://www.seattletimes.com/seattle-news/health/front-line-medical-workers-get-first-doses-of-coronavirus-vaccine-in-seattle/.

Cha, Ariana Eunjung, Brittany Shammas, Ben Guarino, and Jacqueline Dupree. "Record numbers of covid-19 patients push hospitals and staffs to the limit." *The Seattle Times*. December 16, 2020. https://www.seattletimes.com/nation-world/nation/record-numbers-of-covid-19-patients-push-hospitals-and-staffs-to-the-limit/.

Choi, Candace, and Michelle R. Smith. "States ramp up for biggest vaccination effort in US history." *The Seattle Times*. Updated November 13, 2020. https://www.seattletimes.com/seattle-news/health/states-ramp-up-for-biggest-vaccination-effort-in-us-history/.

Cott, Emma, Elliot deBruyn, and Jonathan Corum. "How Pfizer Makes Its Covid-19 Vaccine." *The New York Times*. April 28, 2021. https://www.nytimes.com/interactive/2021/health/pfizer-coronavirus-vaccine.html.

CVS. "CVS Health surpasses 10 million COVID-19 vaccine doses administered." Press release. April 1, 2021. https://cvshealth.com/news-and-insights/press-releases/cvs-health-surpasses-10-million-covid-19-vaccine-doses.

Griffin, Riley. "Pfizer to Deliver U.S. Vaccine Doses Faster Than Expected." *Bloomberg*. January 26, 2021. https://news.bloomberglaw.com/health-law-and-business/pfizer-to-deliver-u-s-vaccine-doses-faster-than-expected-ceo.

Hess, Corrinne. "Pfizer's Pleasant Prairie Facility Could Supply Western US With Coronavirus Vaccine." Wisconsin Public Radio. December 9, 2020. https://www.wpr.org/pfizers-pleasant-prairie-facility-could-supply-western-us-coronavirus-vaccine.

Johnson, Carolyn Y. "A vial, a vaccine and hopes for slowing a pandemic—how a shot comes to be." *The Washington Post*. November 17, 2020. https://www.washingtonpost.com/health/2020/11/17/coronavirus-vaccine-manufacturing/.

Lowe, Derek. "RNA Vaccines And Their Lipids." *In the Pipeline* (blog). *Science Translational Medicine*. January 11, 2021. https://blogs.sciencemag.org/pipeline/archives/2021/01/11/rna-vaccines-and-their-lipids.

Lupkin, Sydney. "Pfizer's Coronavirus Vaccine Supply Contract Excludes Many Taxpayer Protections." NPR. November 24, 2020. https://www.npr.org/sections/health-shots/2020/11/24/938591815/pfizers-coronavirus-vaccine-supply-contract-excludes-many-taxpayer-protections.

Madani, Doha. "First trucks with Covid-19 vaccine roll out of Pfizer plant in Michigan." NBC News. December 13, 2020. https://www.nbcnews.com/news/us-news/first-trucks-covid-19-vaccine-roll-out-pfizer-plant-michigan-n1251037.

Mahase, Elisabeth. "Vaccinating the UK: How the Covid Vaccine Was Approved, and Other Questions Answered." *BMJ* 371 (December 9, 2020): m4759. https://doi.org/10.1136/bmj.m4759.

Mehta, Chavi. "Lyft, CVS Health partner to increase access to COVID-19 vaccines." Reuters. February 19, 2021. https://www.reuters.com/business/healthcare-pharmaceuticals/lyft-cvs-health-partner-increase-access-covid-19-vaccines-2021-02-19/.

Moutinho, Sofia. "Syringe size and supply issues continue to waste COVID-19 vaccine doses in United States." *Science*. March 26, 2021. https://www.sciencemag.org/news/2021/03/syringe-size-and-supply-issues-continue-waste-covid-19-vaccine-doses-united-states.

O'Donnell, Carl. "Why Pfizer's ultra-cold COVID-19 vaccine will not be at the local pharmacy any time soon." Reuters. November 9, 2020. https://www.reuters.com/article/health-coronavirus-vaccines-distribution/why-pfizers-ultra-cold-covid-19-vaccine-will-not-be-at-the-local-pharmacy-any-time-soon-idUSKBN27P2VP.

Park, Alice. "The First Authorized COVID-19 Vaccine in the U.S. Has Arrived." *Time*. December 11, 2020. https://time.com/5920134/first-authorized-covid-19-vaccine-us/.

Pfizer. "COVID-19 Vaccine U.S. Distribution Fact Sheet." November 2020. https://www.pfizer.com/news/hot-topics/covid_19_vaccine_u_s_distribution_fact_sheet.

Pfizer. "Pfizer and BioNTech to Supply the U.S. with 100 Million Additional Doses of COVID-19 Vaccine." Press release. December 3, 2020. https://www.pfizer.com/news/press-release/press-release-detail/pfizer-and-biontech-supply-us-100-million-additional-doses.

Pfizer. "Pfizer Announces Agreement with Gilead to Manufacture Remdesivir for Treatment of COVID-19." Press release. August 7, 2020. https://www.pfizer.com/news/press-release/press-release-detail/pfizer-announces-agreement-gilead-manufacture-remdesivir.

Romo, Vanessa. "Some Vials Of COVID-19 Vaccine Contain Extra Doses, Expanding Supply, FDA Says." NPR. December 16, 2020. https://www.npr.org/sections/coronavirus-live-updates/2020/12/16/947386411/some-vials-of-covid-19-vaccine-contain-extra-doses-expanding-supply.

Routledge. "John D Ludwig." https://www.routledge.com/authors/i3183-john-ludwig#.

Rowland, Christopher. "Biden wants to squeeze an extra shot of vaccine out of every Pfizer vial. It won't be easy." *The Washington Post*. January 22, 2021. https://www.washingtonpost.com/business/2021/01/22/pfizer-vaccine-doses-syringes/.

Rowland, Christopher. "Inside Pfizer's race to produce the world's biggest supply of covid vaccine." *The Washington Post*. June 16, 2021. https://www.washingtonpost.com/business/2021/06/16/pfizer-vaccine-engineers-supply/.

Rowland, Christopher. "Pfizer spent months working to extract sixth dose from vials as vaccine production shortfalls loomed." *The Washington Post*. February 3, 2021. https://www.washingtonpost.com/business/2021/02/03/pfizer-vaccine-syringes-doses/.

Rubicon Science. "Knauer develops Impingement Jet Mixing Technology for the production of mRNA-filled nanoparticles." April 20, 2021. https://rubiconscience.com.au/2021/04/20/knauer-develops-impingement-jets-mixing-technology-for-the-production-of-mrna-filled-nanoparticles/.

Shadburne, Sarah. "UPS ships out first round of Covid-19 vaccines from Louisville." *Louisville Business First.* December 14, 2020. https://www.bizjournals.com/louisville/news/2020/12/14/ups-ships-out-first-round-of-covid-vaccines-from.html.

Stares, Justin. "The beer and beauty fame of the Belgian town that is about to save the world." *Daily Mail.* November 14, 2020. https://www.dailymail.co.uk/news/article-8950061/The-beer-beauty-fame-Belgian-town-save-world.html.

Taliaferro, Lanning. "Rockland Coronavirus: Restrictions Imposed In 4 New Hot Spots." *Patch.* November 19, 2020. https://patch.com/new-york/pearlriver/rockland-coronavirus-restrictions-imposed-4-new-hot-spots.

Thomas, Katie, Sharon LaFraniere, Noah Weiland, Abby Goodnough, and Maggie Haberman. "F.D.A. Clears Pfizer Vaccine, and Millions of Doses Will Be Shipped Right Away." *The New York Times.* December 11, 2020. https://www.nytimes.com/2020/12/11/health/pfizer-vaccine-authorized.html. Excerpt from *The New York Times.* © 2020 The New York Times Company. All rights reserved. Used under license.

Weintraub, Karen, and Elizabeth Weise. "The sprint to create a COVID-19 vaccine started in January. The finish line awaits." *USA Today.* Updated September 12, 2020. https://www.usatoday.com/in-depth/news/health/2020/09/11/covid-vaccine-trials-update-timeline-companies-progress-phases/3399557001/.

World Health Organization. "Tozinameran COVID-19 mRNA Vaccine (nucleoside modified)—COMIRNATY® (Pfizer–BioNTech) Training." January 27, 2021. https://cdn.who.int/media/docs/default-source/immunization/covid-19/pfizer-specific-training_full-deck_27-january-final.pdf.

Chapter 8

Bourla, Albert. "An Open Letter from Pfizer Chairman and CEO to Colleagues." Pfizer. May 7, 2021. https://www.pfizer.com/news/hot-topics/why_pfizer_opposes_the_trips_intellectual_property_waiver_for_covid_19_vaccines.

Britannica. "Ursula von der Leyen." https://www.britannica.com/biography/Ursula-von-der-Leyen.

Centers for Disease Control and Prevention. "CDC Vaccine Price List." Updated July 1, 2021. https://www.cdc.gov/vaccines/programs/vfc/awardees/vaccine-management/price-list/index.html#adult.

Centers for Disease Control and Prevention. "Vaccine Effectiveness: How Well Do the Flu Vaccines Work?" Updated May 6, 2021. https://www.cdc.gov/flu/vaccines-work/vaccineeffect.htm.

Cooper, Ryan. "Trump's jaw-dropping vaccine screwup." *The Week*. December 9, 2020. https://theweek.com/articles/953941/trumps-jawdropping-vaccine-screwup.

Das, Krishna. "India delays big exports of AstraZeneca shot, as infections surge." Reuters. March 24, 2021. https://www.reuters.com/article/health-coronavirus-india-vaccine-exclusi-idINKBN2BG27D.

The Economist. "India's Covid-19 Crisis Has Spiralled out of Control." May 3, 2021. https://www.economist.com/graphic-detail/2021/05/03/indias-covid-19-crisis-has-spiralled-out-of-control.

Gaurav, Kunal. "Covid-19 travel ban: These countries have restricted flights to and from India." *Hindustan Times*. May 4, 2021. https://www.hindustantimes.com/world-news/covid19-travel-ban-these-countries-have-restricted-flights-to-and-from-india-101620125693859.html.

LaFraniere, Sharon, and Zach Montague. "Pfizer Seals Deal With U.S. for 100 Million More Vaccine Doses," *The New York Times*. December 23, 2020. https://www.nytimes.com/2020/12/23/us/politics/pfizer-vaccine-doses-virus.html.

Lupkin, Sydney. "Defense Production Act Speeds Up Vaccine Production." NPR. March 13, 2020. https://www.npr.org/sections/health-shots/2021/03/13/976531488/defense-production-act-speeds-up-vaccine-production.

Lupkin, Sydney. "U.S. Government May Find It Hard To Get More Doses Of Pfizer's COVID-19 Vaccine." NPR. December 10, 2020. https://www.npr.org/sections/health-shots/2020/12/10/944857395/us-government-may-find-it-hard-to-get-more-doses-of-pfizers-covid-19-vaccine.

Office of the United States Trade Representative. "Statement from Ambassador Katherine Tai on the Covid-19 Trips Waiver." Press release. May 5, 2021. https://ustr.gov/about-us/policy-offices/press-office/press-releases/2021/may/statement-ambassador-katherine-tai-covid-19-trips-waiver.

Pfizer. "Angela Hwang." https://www.pfizer.com/people/leadership/executives/angela_hwang.

Pfizer. "Pfizer and BioNTech Celebrate Historic First Authorization in the US of Vaccine to Prevent Covid-19." Press release. December 11, 2020. https://www.pfizer.com/news/press-release/press-release-detail/pfizer-and-biontech-celebrate-historic-first-authorization.

Pfizer. "Pfizer and BioNTech Reach Agreement With Covax For Advance Purchase of Vaccine to Help Combat COVID-19." Press release. January 22, 2021. https://www.pfizer.com/news/press-release/press-release-detail/pfizer-and-biontech-reach-agreement-covax-advance-purchase.

Phartiyal, Sankalp, and Alasdair Pal "India's daily COVID-19 cases pass 400,000 for first time as second wave worsens." Reuters. Updated April 30, 2021. https://www.reuters.com/world/asia-pacific/india-posts-record-daily-rise-covid-19-cases-401993–2021–05–01/.

Reuters. "African Union drops AstraZeneca vaccine, which COVAX will supply." April 8, 2020. https://www.reuters.com/article/uk-health-coronavirus-africa/african-union-drops-astrazeneca-vaccine-which-covax-will-supply-idUSKBN2BV19H.

Reuters. "Pfizer-BioNTech to provide 1 bln vaccines to poorer nations this year." May 21, 2021. https://www.reuters.com/article/health-coronavirus-pfizer-vaccine/rpt-pfizer-biontech-to-provide-1-bln-vaccines-to-poorer-nations-this-year-idUSL5N2N83AQ.

Shear, Michael D., and David E Sanger. "Biden Aims to Bolster U.S. Alliances in Europe, but Challenges Loom." *The New York Times*. Updated June 11, 2021. https://www.nytimes.com/2021/06/09/us/politics/biden-europe-g7.html. Excerpt from *The New York Times*. © 2020 The New York Times Company. All rights reserved. Used under license.

Stacey, Kiran. "Jeff Zients: the 'Mr Fix It' in charge of tackling the US Covid crisis." *Financial Times*, January 20, 2021. https://www.ft.com/content/b52ca23e-d244-498b-8199-5f7bd3f64177.

Stearns, Jonathan, Alberto Nardelli, and Nikos Chrysoloras. "Faced With Vaccine Shortages, EU Set to Impose Export Controls." Bloomberg. January 28, 2021. https://www.bloomberg.com/news/articles/2021–01–28/europe-opens-door-to-vaccine-export-ban-risking-global-backlash.

US Department of Health and Human Services. "Fact Sheet: Explaining Operation Warp Speed." December 21, 2020. https://public3.pagefreezer.com/content/HHS.gov/31-12-2020T08:51/https:/www.hhs.gov/about/news/2020/06/16/fact-sheet-explaining-operation-warp-speed.html.

US Embassy & Consulates in the United Kingdom. "COVID-19 Information." https://uk.usembassy.gov/covid-19-coronavirus-information/.

Twohey, Megan, Ketin Collins, and Katie Thomas. "With First Dibs on Vaccines, Rich Countries Have 'Cleared the Shelves.'" *The New York Times*. Updated December 18, 2020. https://www.nytimes.com/2020/12/15/us/coronavirus-vaccine-doses-reserved.html.

The Week. "Tregenna Castle: inside the Cornwall resort hosting G7 summit leaders." June 10, 2021. https://www.theweek.co.uk/news/uk-news/953111/inside-tregenna-castle-resort-where-world-leaders-will-stay-for-g7.

Weixel, Nathaniel. "US comes under pressure to share vaccines with rest of world." *The Hill*. March 14, 2021. https://thehill.com/policy/healthcare/543004-us-comes-under-pressure-to-share-vaccines-with-rest-of-world.

World Bank. "The World Bank Atlas method—detailed methodology." https://datahelpdesk.worldbank.org/knowledgebase/articles/378832-what-is-the-world-bank-atlas-method.

World Bank. "The World by Income and Region." World Development Indicators. 2020. https://datatopics.worldbank.org/world-development-indicators/the-world-by-income-and-region.html.

World Health Organization. "Biography: Dr Tedros Adhanom Ghebreyesus." https://www.who.int/director-general/biography.

World Health Organization. "Dr Tedros takes office as WHO Director-General." Press release. July 1, 2017. https://www.who.int/news/item/01–07–2017-dr-tedros-takes-office-as-who-director-general.

World Health Organization. "What is the ACT-Accelerator." https://www.who.int/initiatives/act-accelerator/about.

World Health Summit. "Albert Bourla." https://www.worldhealthsummit.org/speaker-view.html?tx_glossary2_glossary%5Bglossary%5D=497&tx_glossary2_glossary%5Baction%5D=show&tx_glossary2_glossary%5Bcontroller%5D=Glossary&cHash=4c5aa69f2ebf2b540822f6a2d525bfd8.

World Trade Organization. "Export Controls and Export Bans over the Course of the Covid-19 Pandemic." Press release. April 29, 2020. https://www.wto.org/english/tratop_e/covid19_e/bdi_covid19_e.pdf.

World Trade Organization. "Overview: the TRIPS Agreement." https://www.wto.org/english/tratop_e/trips_e/intel2_e.htm.

Chapter 9

Aljazeera. "As Olympics begin, Japan rolls out red carpet for Pfizer CEO." July 23, 2021. https://www.aljazeera.com/economy/2021/7/23/japan-rolls-out-red-carpet-for-pfizer-ceo-to-ensure-jab-delivery.

Baron, John. "The Life of Edward Jenner, M.D., LL.D., F.R.S., Physician Extraordinary to His Majesty George IV., Foreign Associate of the National Institute of France, With Illustrations of His Doctrines, and Selections from His Correspondence." *Edinburgh Medical and Surgical Journal* 51, no. 139 (April 1, 1839): 500–527.

Biden, Joseph R. "Inaugural Address by President Joseph R. Biden, Jr." Speech. Washington, DC. January 20, 2021. https://www.whitehouse.gov/briefing-room/speeches-remarks/2021/01/20/inaugural-address-by-president-joseph-r-biden-jr/.

Department of Health and Social Care. "UK COVID-19 vaccines delivery plan." Gov.uk. Updated January 13, 2021. https://www.gov.uk/government/publications/uk-covid-19-vaccines-delivery-plan/uk-covid-19-vaccines-delivery-plan.

Guarascio, Francesco. "EU to shortly sign world's largest vaccine deal with Pfizer." Reuters. April 23, 2021. https://www.reuters.com/business/healthcare-pharmaceuticals/eu-seals-deal-with-pfizer-biontech-up-18-bln-doses-vaccine-eu-official-2021-04-23/.

International Olympic Committee. "Who was Pierre de Coubertin?" April 28, 2021. https://olympics.com/ioc/faq/history-and-origin-of-the-games/who-was-pierre-de-coubertin.

Jenner Institute. "About Edward Jenner." https://www.jenner.ac.uk/about/edward-jenner.

Johnson, Boris. "Excellent conversation with Albert Bourla yesterday." LinkedIn. January 14, 2021. https://www.linkedin.com/posts/boris-johnson_excellent-conversation-with-albert-bourla-activity-6755790234409021440-BfMU/.

Lentz, Thierry. "Talking Point with Thierry Lentz: Vaccination: When Napoleon Declared War on Smallpox." Napoleon.org. https://www.napoleon.org/en/history-of-the-two-empires/articles/talking-point-with-thierry-lentz-vaccination-when-napoleon-declared-war-on-smallpox/.

Maan, Anurag, Shaina Ahluwalia, and Kavya B. "Global coronavirus cases exceed 50 million after 30-day spike." Reuters. November 8, 2020. https://www.reuters.com/article/health-coronavirus-global-cases/global-coronavirus-cases-exceed-50-million-after-30-day-spike-idUSKBN27O0IO.

Navajo Nation OPVP Communications. "Live Town Hall Meeting with Dr. Albert Bourla 12.24.20." YouTube. Streamed live December 24, 2020. https://www.youtube.com/watch?v=wjscKbMSuUk.

Obrador, Andrés Manuel López. "Presidente afianza con Pfizer compromiso de entrega de vacunas." Press release. January 19, 2021. https://lopezobrador.org.mx/2021/01/19/presidente-afianza-con-pfizer-compromiso-de-entrega-de-vacunas/.

Pfizer. "Pfizer and BioNTech to Provide COVID-19 Vaccine Doses for Olympic Athletes at the 2020 Tokyo Games." Press release. May 6, 2021. https://www.pfizer.com/news/press-release/press-release-detail/pfizer-and-biontech-provide-covid-19-vaccine-doses-olympic.

Rickert, Levi. "Navajo Nation President Spoke with Pfizer CEO about Vaccine on Wednesday." Native News Online. December 9, 2020. https://nativenewsonline.net/currents/navajo-nation-president-spoke-with-pfizer-ceo-about-vaccine-on-wednesday.

Shea, Sandra L. "How Can Scientists Promote Peace?" *Temperature* 5, no. 1 (February 22, 2018): 7–8. https://doi.org/10.1080/23328940.2017.1397086.

von der Leyen, Ursula. "Statement by President von der Leyen, Prime Minister of Belgium De Croo, CEO of Pfizer Bourla, and co-founder and Chief Medical Officer of BioNTech Türeci, following the visit to the Pfizer manufacturing plant in Puurs, Belgium." Speech. Puurs, Belgium. April 23, 2021. https://ec.europa.eu/commission/presscorner/detail/en/statement_21_1929.

Zeidler, Maryse. "Canadians' hesitancy about COVID-19 vaccine dropping, new poll suggests." CBC News. March 8, 2021. https://www.cbc.ca/news/canada/british-columbia/vaccine-poll-hesitancy-dropping-1.5940400.

Chapter 10

Arlosoroff, Meirav. "Israel's Population Is Growing at a Dizzying Rate. Is It Up for the Challenge?" *Haaretz*. January 4, 2021. https://www.haaretz.com/israel-news/.premium.MAGAZINE-israel-s-population-is-growing-at-a-dizzying-rate-is-it-up-for-the-challenge-1.9410043.

Benmeleh, Yaacov. "World-Leading Vaccine Push Augurs Return to Normal in Israel." Bloomberg. February 16, 2021. https://www.bloomberg.com/news/articles/2021-02-16/world-s-fastest-vaccine-push-augurs-return-to-normal-in-israel.

Bourla, Albert. "While Caution Should Be Used in Extrapolating to Other Countries, These Observational Findings Are Demonstrating the Impact We Have in Reducing Human Pain and Makes All of Us at Pfizer so Proud: https://Bit.Ly/3wcS6qx (2/2)." Twitter, May 17, 2021. https://twitter.com/AlbertBourla/status/1394281683009089544.

Centers for Disease Control and Prevention. "Joint CDC and FDA Statement on Vaccine Boosters." Press statement. July 8, 2021. https://www.cdc.gov/media/releases/2021/s-07082021.html.

Centers for Disease Control and Prevention. "Joint Statement from HHS Public Health and Medical Experts on COVID-19 Booster Shots." Press statement. August 18, 2021. https://www.cdc.gov/media/releases/2021/s0818-covid-19-booster-shots.html.

Centers for Disease Control and Prevention. "Media Statement from CDC Director Rochelle P. Walensky, MD, MPH, on Signing the Advisory Committee on Immunization Practices' Recommendation for an Additional Dose of an mRNA COVID-19 Vaccine in Moderately to Severely Immunocompromised People." Press statement. August 13, 2021. https://www.cdc.gov/media/releases/2021/s0813-additional-mRNA-mrna-dose.html.

Centers for Disease Control and Prevention. "Science Brief: Background Rationale and Evidence for Public Health Recommendations for Fully Vaccinated People." Updated April 2, 2021. https://stacks.cdc.gov/view/cdc/104739.

Centers for Disease Control and Prevention. "Science Brief: COVID-19 Vaccines and Vaccination." Updated July 27, 2021. https://www.cdc.gov/coronavirus/2019-ncov/science/science-briefs/fully-vaccinated-people.html.

Dimolfetta, David. "Netanyahu visits grave of Yoni, killed during Operation Entebbe 43 years ago." *The Jerusalem Post*. July 10, 2019. https://www.jpost.com/israel-news/netanyahu-visits-grave-of-murdered-brother-commemorating-43-years-595197.

Erman, Michael, and Maayan Lubell. "Pfizer/BioNTech say data suggests vaccine 94% effective in preventing asymptomatic infection." Reuters. March 11, 2021. https://www.reuters.com/article/us-health-coronavirus-pfizer-israel-idUSKBN2B31IJ.

Friedman, Gabe. "Meet Mikael Dolsten, the Jewish immigrant leading Pfizer's vaccine charge." *Baltimore Jewish Times*. December 30, 2020. https://www.jewishtimes.com/meet-mikael-dolsten-the-jewish-immigrant-leading-pfizers-vaccine-charge/.

Haas, Eric J., Frederick J. Angulo, John M. McLaughlin, Emilia Anis, Shepherd R. Singer, Farid Khan, Nati Brooks, et al. "Impact and Effectiveness of MRNA BNT162b2 Vaccine against SARS-CoV-2 Infections and COVID-19 Cases, Hospitalisations, and Deaths Following a Nationwide Vaccination Campaign in

Israel: An Observational Study Using National Surveillance Data." *The Lancet* 397, no. 10287 (May 15, 2021): 1819–29. https://www.thelancet.com/journals/lancet/article/PIIS0140-6736(21)00947-8/fulltext. Excerpt reprinted from *The Lancet*, copyright © 2021, with permission from Elsevier.

Jaffe-Hoffman, Maayan. "Israel signs agreement with Pfizer, Moderna for millions more COVID-19 vaccines." *The Jerusalem Post*. April 20, 2021. https://www.jpost.com/health-science/israel-to-sign-agreement-for-millions-more-pfizer-vaccines-665621.

Lardieri, Alexa. "All Israelis Over 16 Are Eligible for Coronavirus Vaccine." *U.S. News & World Report*. February 4, 2021. https://www.usnews.com/news/health-news/articles/2021-02-04/all-israelis-over-16-are-eligible-for-coronavirus-vaccine.

Pfizer. "Pfizer and BioNTech Provide Update on Booster Program in Light of the Delta-Variant." Press release. July 8, 2021. https://cdn.pfizer.com/pfizercom/2021-07/Delta_Variant_Study_Press_Statement_Final_7.8.21.pdf.

Ritchie, Hannah, Esteban Ortiz-Ospina, Diana Beltekian, Edouard Mathieu, Joe Hasell, Bobbie MacDonald, Charlie Giattino, Cameron Appel, Lucas Rodes-Guirao, and Max Roser. "Coronavirus Pandemic (COVID-19)." OurWorldInData.org. 2020. https://ourworldindata.org/coronavirus. Licensed under CC BY 4.0.

Srivastava, Mehul. "Israelis raise glass to Pfizer as lockdown ends." *Financial Times*. March 12, 2021. https://www.ft.com/content/4cf1b235-ed07-4ffe-bab4-95846a0ecf36.

State of Israel Ministry of Health. "MoH Pfizer Collaboration Agreement." January 6, 2021. https://govextra.gov.il/media/30806/11221-moh-pfizer-collaboration-agreement-redacted.pdf.

Tikkanen, Roosa, Robin Osborn, Elias Mossialos, Ana Djordjevic, and George A. Wharton. "Israel." The Commonwealth Fund: International Health Care System Profiles. June 5, 2020. https://www.commonwealthfund.org/international-health-policy-center/countries/israel.

Times of Israel. "Pfizer CEO hails 'obsessive' Netanyahu for calling 30 times to seal vaccine deal." March 11, 2021. https://www.timesofisrael.com/pfizer-ceo-obsessive-netanyahu-called-30-times-in-effort-to-seal-vaccine-deal/.

Times of Israel. "Pfizer CEO to Visit Israel in March—Report." February 21, 2021. https://www.timesofisrael.com/liveblog_entry/pfizer-ceo-to-visit-israel-in-march-report/.

VOA News. "World Marks One-Year Anniversary of WHO's Official Declaration of COVID-19 Pandemic." March 11, 2021. https://www.voanews.com/covid-19-pandemic/world-marks-one-year-anniversary-whos-official-declaration-covid-19-pandemic.

Chapter 11

Appeal of Conscience Foundation. "Dr. Albert Bourla acceptance remarks upon receiving the 2021 Appeal of Conscience Award." May 26, 2021. https://appeal

ofconscience.org/dr-albert-bourla-acceptance-remarks-upon-receiving-the-2021-appeal-of-conscience-award/.

Bourla, Albert. "An Open Letter from Pfizer Chairman and CEO Albert Bourla." Pfizer. October 16, 2020. https://www.pfizer.com/news/hot-topics/an_open_letter_from_pfizer_chairman_and_ceo_albert_bourla.

Doshi, Peter. "Covid-19 Vaccine Trial Protocols Released." *BMJ* 371 (October 21, 2020): m4058. https://doi.org/10.1136/bmj.m4058.

Gottfried, Jeffrey, Mason Walker, and Amy Mitchell. "Americans' Views of the News Media During the COVID-19 Outbreak." Pew Research Center. May 8, 2020. https://www.journalism.org/2020/05/08/americans-views-of-the-news-media-during-the-covid-19-outbreak/.

Lee, Bruce Y. "Trump Suggests 'Deep State' At FDA Is Delaying Covid-19 Coronavirus Vaccine Testing." *Forbes*. August 22, 2020. https://www.forbes.com/sites/brucelee/2020/08/22/trump-says-deep-state-or-whoever-at-fda-delaying-covid-19-coronavirus-vaccine-testing/?sh=632e53b0f48d.

Merriam-Webster. "Trust." https://www.merriam-webster.com/dictionary/trust.

Pfizer. "Biopharma Leaders Unite To Stand With Science." Press release. September 8, 2020. https://www.pfizer.com/news/press-release/press-release-detail/biopharma-leaders-unite-stand-science.

Pfizer. "Let's Undo Underrepresented Diversity in Clinical Trials." https://www.pfizer.com/science/clinical-trials/diversity-clinical-trials.

Pfizer. "No Stone Left Unturned In The Fight Against COVID-19." YouTube. September 2, 2020. https://www.youtube.com/watch?v=_PBLoSN7OUo.

Pfizer. "Science Will Win-Ask Science." YouTube. April 15, 2020. https://www.youtube.com/watch?v=X10tEfLve1U.

Robert Wood Johnson Foundation and the Harvard T.H. Chan School of Public Health. "The Public's Perspective on the United States Public Health System." May 13, 2021. https://www.rwjf.org/en/library/research/2021/05/the-publics-perspective-on-the-united-states-public-health-system.html.

Snyder Bulik, Beth. "Pharma's reputation rehab: A whopping two-thirds of Americans now offer a thumbs-up, Harris Poll finds." *Fierce Pharma*. February 19, 2021. https://www.fiercepharma.com/marketing/pharma-reputation-hits-high-americans-two-thirds-now-give-positive-rating-harris-poll.

Stacey, Kiran. "FDA head says he is willing to fast-track Covid-19 vaccine." *Financial Times*. August 30, 2020. https://www.ft.com/content/f8ecf7b5-f8d2-4726-ba3f-233b8497b91a.

Sweet, Jesse, dir. "Mission Possible: The Race for a Vaccine." National Geographic CreativeWorks: Washington, DC, 2021. Aired March 11, 2021, on National Geographic. https://www.youtube.com/watch?v=jbZUZ9JYNBE.

Vardi, Nathan. "The Race Is On: Why Pfizer May Be the Best Bet to Deliver a Vaccine by the Fall." *Forbes*. May 20, 2020. https://www.forbes.com/sites/nathanvardi/2020/05/20/the-man-betting-1-billion-that-pfizer-can-deliver-a-vaccine-by-this-fall/?sh=385a522382e9.

VOA News. "Fauci 'Cautiously Optimistic' About Coronavirus Vaccine." July 31, 2020. https://www.voanews.com/covid-19-pandemic/fauci-cautiously-optimistic-about-coronavirus-vaccine.

Wagner, Meg, and Melissa Macaya. "Fauci testifies on coronavirus response." CNN. July 31, 2020. https://www.cnn.com/politics/live-news/fauci-coronavirus-testimony-07–31–20/h_d880e3e2e3cedbfdce59805138f3477b.

Westall, Mark. "Pfizer's Vaccine Branding Victory Delivers Lessons for Marketers Everywhere." *AdAge*. July 19, 2021. https://adage.com/article/opinion/pfizers-vaccine-branding-victory-delivers-lessons-marketers-everywhere-opinion/2351431.

Chapter 12

Cubanski, Juliette, Tricia Neuman, Kendal Orgera, and Anthony Damico. *No Limit: Medicare Part D Enrollees Exposed to High Out-of-Pocket Drug Costs Without a Hard Cap on Spending*. The Henry J. Kaiser Family Foundation. November 2017. https://files.kff.org/attachment/Issue-Brief-No-Limit-Medicare-Part-D-Enrollees-Exposed-to-High-Out-of-Pocket-Drug-Costs-Without-a-Hard-Cap-on-Spending.

European Commission. "Health technology assessment." https://ec.europa.eu/health/technology_assessment/overview_en.

Miao, Lei, Yu Zhang and Leaf Huang. "mRNA vaccine for cancer immunotherapy." *Molecular Cancer* 20, no. 41 (February 25, 2021). https://molecular-cancer.biomedcentral.com/articles/10.1186/s12943-021-01335-5.

Pfizer. "Pfizer Reports Strong First-Quarter 2021 Results." Press release. May 4, 2021. https://investors.pfizer.com/investor-news/press-release-details/2021/PFIZER-REPORTS-STRONG-FIRST-QUARTER-2021-RESULTS/default.aspx.

Pfizer. "Rod MacKenzie, PH.D." https://www.pfizer.com/people/leadership/executives/rod_mackenzie-phd.

Smithsonian National Air and Space Museum. "The Journey Home." Apollo to the Moon. https://airandspace.si.edu/exhibitions/apollo-to-the-moon/online/apollo-11/journey-home.cfm.

The White House. "Fact Sheet: President Biden Takes Executive Actions to Tackle the Climate Crisis at Home and Abroad, Create Jobs, and Restore Scientific Integrity Across Federal Government." Press release. January 7, 2021. https://www.whitehouse.gov/briefing-room/statements-releases/2021/01/27/fact-sheet-president-biden-takes-executive-actions-to-tackle-the-climate-crisis-at-home-and-abroad-create-jobs-and-restore-scientific-integrity-across-federal-government/.

World Health Organization. "Cancer." March 3, 2021. https://www.who.int/news-room/fact-sheets/detail/cancer.

Epilogue

The Carter Center. "Corporate, Government, and Foundation Giving." https://www.cartercenter.org/donate/corporate-government-foundation-partners/index.html.

The Carter Center. "Jimmy Carter (biography)." November 6, 2019. https://www.cartercenter.org/about/experts/jimmy_carter.html.

Pfizer. "Board Members." https://www.pfizer.com/people/leadership/board-of-directors.

Weiner, Stacy. "Applications to medical school are at an all-time high. What does this mean for applicants and schools?" American Association of Medical Colleges (AAMC). October 22, 2020. https://www.aamc.org/news-insights/applications-medical-school-are-all-time-high-what-does-mean-applicants-and-schools.

About the Author

During more than twenty-five years at Pfizer, Dr. Albert Bourla built a diverse and successful career, holding a number of senior global positions across a range of countries and disciplines. He moved to Pfizer from academia, and Pfizer is the only corporation he ever worked for. Together with his wife, Myriam, and children, Mois and Selise, they lived in eight different cities of five different countries.

As chairman and chief executive officer, he focused the company on a new purpose: "Breakthroughs that change patients' lives," with a concentrated emphasis on driving the scientific and commercial innovation needed for transformational impacts on human health. Prior to taking the reins as CEO in January 2019, he served as Pfizer's Chief Operating Officer beginning in January 2018, responsible for overseeing the company's commercial strategy, manufacturing, and global product development functions. Previously, from February 2016 to December 2017, he served as Group President of Pfizer Innovative Health and broke up the company into six business units that started operating as "independent" biotechs, competing for resources on the merits of their breakthrough research projects. From January 2014 to January 2016, he was Group President of Pfizer's Global Vaccines, Oncology, and Consumer Healthcare businesses. Dr. Bourla started at Pfizer in 1993 in the Animal Health division as technical director of Greece. He held positions of increasing responsibility within the Animal Health division across Europe before moving to Pfizer's New York global headquarters in 2001.

He is a doctor of veterinary medicine and holds a PhD in the biotechnology of reproduction from the veterinary school of Aristotle University. In 2020, having completed one year as CEO, he was ranked as America's top CEO in the pharmaceuticals sector by *Institutional Investor* magazine. In 2021, he received the Appeal of Conscience Award from the homonymous foundation; the David Rockefeller Award from the Museum of Modern Art; the Father of the Year Award from the National Father's Day Committee, benefiting the Save the Children organization; the Athenagoras Human Rights Award from the Order of Saint Andrew (Archons of the Ecumenical Patriarchate in America); the Roy Vagelos Pro Bono Humanum Award for Global Health Equity from the Galien Foundation; the Theodor Herzl Award from the World Jewish Congress; and the Distinguished Leadership Award from the Atlantic Council, among others. He has also received the Grand Cordon of the Order of Independence from King Abdullah II of Jordan and the Gold Cross of the Order of the Redeemer, from President Sakellaropoulou of Greece.

Dr. Bourla holds honorary degrees from the Medical School of Aristotle University of Thessaloníki and Babson College. He is on the executive committee of the Partnership for New York City, a director on multiple boards—Pfizer Inc., PhRMA, and Catalyst—and a trustee of the United States Council for International Business.